ILLUSAFACT . . .
THE INEVITABLE ADVANCE
OF OUR TECHNOLOGIES
AND US

ILLUSAFACT . . .
THE INEVITABLE ADVANCE
OF OUR TECHNOLOGIES
AND US

Dr. Robert H. Schram

To order additional copies of this book, contact:
Xlibris Corporation
1-888-795-4274
www.Xlibris.com
Orders@Xlibris.com
95595

Contents

INTRODUCTION

The collective quantity of information in the world is beyond our comprehension. It is growing at an incredible rate (a compound annual 60% growth) that is speeding up all the time. Most significantly, for the purpose of this book is that information created by machines and used by other machines is growing faster than any other form of data. Humans are only tangentially involved in this primarily 'database to database' information. No one can deny that the only way to efficiently access this gargantuan quantity of data is by utilizing machines. Human minds can handle, on average, seven pieces of information in our short-term memory and can generally deal with only four concepts or relationships at once. People become confused when there is more information to process, or it is especially complex.

In medicine the concept of physicians trying to keep up with their entire industry was given up many decades ago since today they would need to be familiar with about 10,000 diseases, 3,000 drugs and more than 1,000 lab tests. Computers aggregate and winnow raw information into a processed format so people can use it. The use of "predictive analytics" on the basis of large data sets is transforming health care. With premature babies International Business Machine (IBM) has developed a system that monitors subtle changes in seven streams of real-time data, such as respiration, heart rate and blood pressure. The electrocardiogram alone generates 1,000 readings per second. This kind of information is turned out by all medical equipment, but previously it was recorded on paper and examined perhaps once an hour. By feeding the data into a computer, the onset of an infection can be detected before obvious symptoms emerge. Analytics, recursive machine-learning, and visualization software is used to make data more digestible for human consumption. Today mathematical algorithms are doing more of the thinking for people and this clearly has significant risks. For every success with big data there are many failures: 1. Banks were unable to understand their risks in the lead-up to the financial melt-down in 2008; 2. On May 7, 2010 the Dow Jones Average abruptly dropped 1,000 points in twenty minutes; 3. The inadequate system used to identify potential terrorists (Times Square bomber and known terrorists boarding airplanes) and secure secret databases (America's National Security Agency (NSA) in January 2000 crashed because of the enormous amount of data pouring into its data banks; for 3.5 days the agency was unable to process information compromising America's safety).

After 9-11-01 the United States Defense Department launched "Total Information Awareness" to compile as much information as possible on just about everything: e-mails, phone calls, web searches, shopping transactions, bank records, medical files, travel history, and much more. Since 1996 "Internet Archive" has been recording all the content on the web as a not-for-profit venture; it now includes software, films, audio recordings, and scanned books. In commercial arenas across the Internet economy, companies are compiling masses of data on people, their activities, their likes and dislikes, their relationships with others and even where they are at any particular moment. Information is accumulated surreptitiously for profit motives. Amazon not only tracks the books you purchase, but also keeps a record of the ones you browse but do not buy to help it recommend other books to you. Information from its e-book, the Kindle, probably records how long a user spends reading each page, whether s/he takes notes, etc. Facebook, a social-networking site, tracks the activities of its 500 million plus and growing users, half of whom spend an average of almost an hour on the site every day. Internet businesses are uncomfortable with full transparency and disclosure because it is at the heart of their competitive advantage.

Google is the king of the gargantuan data collectors with its audacious mission to "organize the world's information." The lifeblood of Google is creating new economic value from unthinkably large amounts of information. Google exploits information that is a by-product of user interactions, or data exhaust, which is automatically recycled to improve the service or create an entirely new product. Translation and voice recognition are two of Google's newest and growing services. Google Translate now covers more than 50 languages. The system identifies which word or phrase in one language is the most likely equivalent in a second language.

Analyzing and growing data bases are ripe with challenges since most chief information officers (CIO's) admit that their data is not of high quality. Many say that the technology meant to make sense of the data often just produces more data. Instead of finding a needle in the haystack, they are making more hay. Business decisions nevertheless are increasingly being made or at least corroborated on the basis of computer algorithms rather than on individual intuition or hunches. Wal-Mart is a good example of the need to use analytic technology. The retailer operates 8,400 stores worldwide, has more than two million employees and handles over 200 million customer transactions each week. Its revenue in 2009, around $400 billion, is more than the GDP of many countries. Wal-Mart's inventory-management system, called Retail Link, enables suppliers to see the exact number of their products on every shelf of every store at that precise moment. The system shows the rate of sales by the hour, by the day, over the past year, and more.

Only 5% of the information that is created is "structured," meaning it comes in a standard format of words or numbers that can be read by computers. The rest are things like photos and phone calls which are less easily retrievable and usable. But this is changing as content on the web is increasingly "tagged," and facial-recognition and voice-recognition software can identify people and words in digital files.

Humanity's dependence on machines has been with us since the Industrial Revolution over 100 years ago but in the 21st century our dependence has developed into a total reliance. This is not to say that Arnold Schwarzenegger's character in "The Terminator" is coming any time soon. We have already been very significantly impacted by 'the rise of the machines' in every venue of human endeavor. Some impacts have had mostly positive results (medicine, discovery, exploration) while others have had serious negative results (economics, safety, homeland security). As we continue to enhance artificial intelligence and more and more distance ourselves from exactly how the software/hardware works we will continue to methodically evolve into losing control of our creations. Dependence leads to more reliance and more reliance leads to more confidence and more confidence leads to less control. The world economy has already been rocked by this irresponsible confidence combined with irrational exuberance and greed. More and more accidents are being caused by machine failures (automobile recalls, airline crashes, civilian deaths in Iraq and Afghanistan). Since the world-wide economic meltdown in 2008 no meaningful change has been enacted to prevent its reoccurrence. As a species we have made tremendous progress in science and improving the quality of our lives but our nature has not changed one iota (complex lovers and killers, and all that goes in between). 'Business as usual' seems to have always been the way civilizations have operated . . . it just seems that the stakes are much higher now!

As information becomes more abundant, the main problem is no longer finding the information but laying one's hands on the relevant bits easily and quickly. What is needed is information about information. Librarians and computer scientists call it metadata. Information management has a long history. In Assyria around three millennia ago clay tablets had small clay labels attached to them to make them easier to tell apart when they were filed in baskets or on shelves. The idea survived into the 20th century in the shape of the little catalogue cards librarians used to note a book's title, author, subject and so on before the records were moved onto computers. The actual books constituted the data, the catalogue cards the metadata. Other examples include package labels of the five billion bar codes that are scanned throughout the world every day.

In the 21st century metadata is undergoing a virtual renaissance. In order to be useful, the cornucopia of information provided by the Internet has to be organized. That is what Google does so well. The raw material for its search engines comes free: web pages on the public Internet. Where it adds value (and creates metadata) is by structuring the information and ranking it in order of its relevance to the query. Google handles around half the world's Internet searches, answers around 35,000 queries every second. Metadata is a potentially lucrative business. "If you can control the pathways and means of finding information, you can extract rents from subsequent levels of producers," explains Eli Noam, a telecoms economist at New York's Columbia Business School. But there are more benign uses too. For example, photos uploaded to the website Flickr contain metadata such as when, how often, and where they were snapped, as well as the camera model . . . useful for would-be buyers.

Internet users help to label unstructured information so it can be easily found, tagging photos and videos. But they have disdained conventional library classifications. Instead, they pick any word they fancy and create an eclectic "folksonomy." So instead of labeling a photograph of Barack Obama as "President," they might call it "sexy" or "SOB." That sounds chaotic, but needn't be. When information was recorded on a tangible medium-paper, film and so on – everything had only one correct place. With digital information the same item can be filed in several places at once, notes David Weinberger, the author of a book about taxonomy and the Internet, "Everything Is Miscellaneous." Digital metadata make things more complicated and simpler at the same time.[1]

The exact amount of information in the world is not known but is clearly monstrous and growing at a very fast rate. In 2007 the flood of data from all our electronic devices (sensors, computers, research labs, cameras, phones, etc.) surpassed the capacity of our storage technologies. Europe's particle-physics laboratory near Geneva (the large Hadron Collider at Cern) generates 40 terabytes every second . . . way beyond our capacity to store and analyze. The scientists currently collect what they can and let all the rest evaporate into the ether. According to a 2008 study by International Data Corp (IDC), a market-research firm, around 1,200 exabytes of digital data would be generated in 2008. The chart below clearly illustrates our "beyond human comprehension" journey with data inflation:

DATA INFLATION		
UNIT	SIZE	WHAT IT MEANS
Bit (b)	1 or 0	Short for 'binary digit,' after the binary code (1 or 0) computers use to store and process data
Byte (B)	8 bits	Basic unit of computing . . . enough to form a letter or a number in computer code
Kilobyte (KB)	1,000 or 2^{10} bytes	From '1,000' in Greek . . . one page of typed text is 2KB
Megabyte (MG)	1,000KB; 2^{20} bytes	From 'large' in Greek . . . complete works of Shakespeare total 5MB
Gigabyte (GB)	1,000MB 2^{30} bytes	From 'giant' in Greek . . . A two hour film can be compressed into 1-2 GB
Terabyte (TB)	1,000GB 2^{40} bytes	From 'monster' in Greek . . . All the books in the U.S. Library of Congress total 15TB
Petabyte (PB)	1,000TB 2^{50} bytes	All letters from U.S. postal service in one year total about to 5PB. Google processes around 1PB each hour.
Exabyte (EB)	1,000PB 2^{60} bytes	Equivalent to 10 billion copies of the magazine "The Economist"
Zettabyte (ZB)	1,000EB 2^{70} bytes	The total amount of information in existence in 2010 is estimated to be about 1.27ZB
Yottabyte (YB)	1,000ZB 2^{80} bytes	Currently too large to imagine

Source "The Economist"

Not all information is consumed by our species but the amount that is consumed is staggering. Researchers at the University of California in San Diego (UCSD) examined the flow of data to American households in 2008 finding that such households were bombarded with 3.6 zettabytes of information (or 34 gigabytes per person per day). The biggest data hogs were video games and television. In terms of bytes, written words are insignificant, amounting to less than 0.1% of the total. However, the amount of reading people do, previously in decline because of television, has almost tripled since 1980, thanks to all that text on the Internet. In the past information consumption was largely passive, leaving aside the telephone. UCSD claims that half of all bytes are received interactively. Shockingly only 5% of the information created is 'structured' (i.e., it comes in a standard format of words or numbers that can be read by computers). Unstructured information is things like photos and phone calls which are less easily retrievable and usable. As content on the Web is increasingly "tagged," and facial-recognition plus voice-recognition software can identify people and words in digital files this unstructured information is rapidly becoming more structured. Significantly, "information created by machines and used by other machines will probably grow faster than anything else," explains Roger Bohn of the UCSD, one of the authors of the study on American households. "This is primarily 'database to database' information – people are only tangentially involved in most of it."[2]

When dealing with solar systems and galaxies one would expect that the information collected would approach ever increasing astronomical levels. In 2000 the Sloan Digital Sky Survey telescope began to collect information in New Mexico, U.S.A. In the first few weeks more data was collected than in the entire history of astronomy. In 2010 its archive contained 140 terabytes of information. The Large Synoptic

Survey Telescope in Chile, due to begin in 2016, will acquire 140 terabytes of information every five days.

Closer to home (i.e., on our planet Earth) astronomical amounts of information is also being collected. The retail giant Wal-Mart handles more than 1 million customer transactions every hour, feeding databases estimated at more than 2.5 petabytes – the equivalent of 167 times the books in America's Library of Congress. Facebook, the social-networking website, is home to forty billion photographs. Decoding the human genome took ten years in 2003 (analyzing three billion base pairs). In 2010 analyzing three billion base pairs could be achieved in one week.

The world contains an unimaginably vast amount of digital information which is getting larger and larger at ever increasing speeds. Properly managed this information can spot business trends, prevent diseases, fight crime, etc. Managed well it can unlock new sources of economic value, provide new scientific insights, and hold governments to account. This ever increasing gargantuan data is creating new problems: the information already exceeds the available storage space and is forecast to become exponentially even more excessive. There is an abundance of tools to capture, process and share all this information (sensors, computers, mobile phones, etc.) but ensuring data security and protecting privacy is becoming much more difficult as the information multiplies and is shared ever more widely around the world.[3]

My interest in writing this book came from my own personal experiences with rapid innovations in our digital world starting with desktop computers, then laptop computers, then cellular telephones, then smart phones, then Ipods, then Geographic Positioning Satellites (GPS), and voice recognition software. The information that is being expanded daily cannot be easily comprehended as we continually develop more and better ways to access, store, and interpret it. The book is divided into nine chapters: Prophetic or Insane? Governance and Democracy, Technology, Hollywood and Science Fiction, Proliferation, Economics and Business, Artificial Intelligence, The Industrial Revolution and the Luddites, and Summary and Conclusions. Read, wonder, and most of all enjoy!

REFERENCES INTRODUCTION

1. *The Economist* (February 25, 2010). "Needle in a Haystack."
2. Bohn, Roger E., Short, James E. (December, 2009). "How Much Information? 2009 Report on American Consumers." *Global Information Industry Center University of California, San Diego.*
3. *The Economist* (February, 25, 2010). "Data, Data Everywhere."

PROPHETIC OR INSANE?

The serial killer Theodore Kaczynski in the 1990's wrote a manifesto infamously titled: "The Unabomber Manifesto" which was published by The New York Times and The Washington Post in cooperation with the FBI in the hope that someone would recognize his anti-machine ranting thereby assisting in his apprehension. His brother did recognize the writing and helped the FBI locate and arrest his errant sibling. In his seventeen years of bombings that killed three people and injured twenty-nine he eluded all attempts at capture. Kaczynski was the Chicago-born son of a Polish sausage maker and a mother who dedicated herself to cultivating the apparent early genius of her first born child. Kaczynski was a sixteen-year-old scholarship student at Harvard who lived in the service quarters of Eliot House and was known to argue Kant in an all-night cafeteria. He went on to graduate school at Michigan and worked in a field of calculus so outside the mainstream that he was advised in the interests of a career to abandon it. He stubbornly refused and when he received his doctorate he was hired by the Mathematics Department at Berkeley where the chairperson of the department judged him to be "probably one of the top twenty to twenty-five PhDs out of eight hundred coming out that year."

At the age of twenty-five in 1967 he was an assistant professor at Berkeley, one of the few in a high-powered department believed to be on an assured tenure track. In 1969 amid riots on Telegraph Avenue, People's Park, and the presence of the National Guard, he abruptly left both Berkeley and academic life despite his department's attempts to keep him.

For a period he supported himself in Salt Lake doing odd jobs before purchasing a small Montana plot (1.5 acres) with his younger brother David. The plot was located four miles outside Lincoln and seven hundred yards from an operating sawmill, to which he would venture sporadically for the rest of his life as a free man. He took bus trips: Lincoln to Helena, Helena to Butte or Missoula for Salt Lake, connected out of Salt Lake for Sacramento or San Francisco. In Helena he was remembered to have bought and sold used books at a local shop, Aunt Bonnie's. In Sacramento he was remembered by the clerks at Tower Books, twenty blocks from the bus station. The clerks referred to him as "Einstein."

His Manifesto, was a typewritten thirty-five-thousand-word manuscript mailed in June 1995 to The New York Times and The Washington Post. On the strength of the writer's promise to "desist from all terrorist activities" if either paper ran the full text within three months, the

Manifesto was published, in September 1995, by both papers, and a month later by a small Berkeley publisher who immediately moved a five-thousand-copy first printing onto the San Francisco Chronicle's bestseller list. Until the publication of the Manifesto, there had been only a few things about the writer we knew, or thought we knew. The pattern of the bombs and their postmarks had suggested familiarity with northern California; an earlier communiqué had even mentioned testing devices in "the sierras," which was not common usage in the country at large but was how people in and around Sacramento refer to the Sierra Nevada.

The targets themselves (academics in the sciences, computer experts, lobbyists for the logging industry) had strongly suggested someone bent on taking to its diehard conclusion a kind of romantic environmentalism that had flourished in northern California during the late 1960s and the 1970s. The public erroneously thought that the bomber, when captured would be a dug-in survivor of one or another of the radical underground groups common during those turbulent times.

At first glance, the Manifesto, which was called 'Industrial Society and Its Future,' seemed to support the notion that the bomber was a diehard environmentalist. The general thesis of the manifesto echoed the apprehension of technology as a double-edged sword that pervaded a good deal of nineteenth-century social thinking. Its central argument was that the consequences of the Industrial Revolution, even though it has "greatly increased the life-expectancy of those of us who live in 'advanced' countries," has also "destabilized society," "made life unfulfilling," "subjected human beings to indignities," "led to widespread psychological suffering," and "inflicted severe damage on the natural world."

The document was logically written but like the Unabomber's previous communiqués, purported to be the product of an unknown underground group, "FC," for "Freedom Club." To the reader the changes the writer had in mind seemed to suggest that they could occur in a kind of geologic time, history's great clock moving into another inexorable correction. A "revolution against the industrial system" was definitely advocated, but it did not need to be violent ("This revolution may or may not make use of violence") or immediate: "It may be sudden or it may be a relatively gradual process spanning a few decades. We can't predict any of that."

The writer seemed modest to a fault, apologetic about his inadequate ability not only to predict all outcomes but to explain all terms, weave together all threads. Again and again, he did "not pretend" to offer "an accurate description," only "a rough indication." Consistently, he acknowledged the "many objections" that could be raised. Repeatedly, he fretted that the principles he presented were "expressed in imprecise language." He ventured to present them all "not as inviolable laws but as rules of thumb, or guides to thinking." He recognized that his discussion had "a serious weakness," regretted that it must remain "far from clear." "Throughout this article," he concluded, ". . . we've made imprecise statements and statements that ought to have had all sorts of qualifications and reservations attached to them; and some of our statements may be flatly false . . . . In a discussion of this kind one must rely heavily on intuitive judgment, and that can sometimes be wrong . . . . We don't claim that this article expresses more than a crude approximation to the truth . . . ."

Theodore Kaczynski was judged legally competent to stand trial and pled guilty in the Sacramento federal court to the charges relating to five bombings and admitted to the remaining eleven. The question of whether or not he was also insane dominated the aborted trial and the national discussion of the case ever since. David Gelernter, the associate professor of computer science at Yale who in 1993 was severely injured and permanently maimed by one of the bombs became a reliable quote on the subject of Kaczynski's insanity by citing: "our 'don't be judgmental' perversity," our "morally disastrous unwillingness to draw a sharp, hard line between good and evil," and in other words, our "moral depravity." "The twentieth century is the crime scene," Gelernter declared, and the blast that injured him had been "a reenactment of a far bigger one a generation earlier, which destroyed something basic in this society that has yet to be repaired." The "tendency among some intellectuals and journalists to dignify with analysis the thinking of violent criminals has always struck me as low and contemptible," he wrote in TIME, "apparently geared up by the crusade to pretty much jettison that part of the canon that had taken as its subject the mysteries of crime and punishment." To call the Unabomber mad, as TIME had done when it referred to Kaczynski as a "mad genius," Gelernter wrote in "Drawing Life: Surviving the Unabomber," "went beyond funny and obnoxious into the realm of evil."

In reading the full text of the Manifesto it was quite obvious to many readers that the author was past the point of no return. Dr. Kaczynski was obsessive about shaping the past quarter century to fit his thesis as if transmitted by a higher power. The text made clear that he was neither lined up with the radical left or the radical right. At times he shared views with neither, or both, or with no one at all. "The conservatives are fools," he advised us. "They whine about the decay of traditional values, yet they enthusiastically support technological progress and economic growth. Apparently it never occurs to them that you can't make rapid, drastic changes in the technology and the economy of a society without causing rapid changes in all other aspects of the society as well, and that such rapid changes inevitably break down traditional values." If he held "the conservatives" in contempt, he also despised "leftists," by which he meant "mainly socialists, collectivists, 'politically correct' types, feminists, gay and disability activists, animal rights activists and the like," an assortment he collectively dismissed as "one of the most widespread manifestations of the craziness of our world." Leftists, he wrote . . . hate anything that has an image of being strong, good and successful. They hate America, they hate Western civilization, they hate white males, they hate rationality.

He claimed the reasons that leftists give for hating the West, etc. do not correspond with their real motives. They say they hate the West because it is warlike, imperialistic, sexist, ethnocentric, and so forth, but where these same faults appear in socialist countries or in primitive countries, the leftist finds excuses for them . . . it is clear that these faults are not the leftist's real motive for hating America and the West. They hate America and the West because they are strong and successful . . . . Words like "self-confidence," "self-reliance," "initiative," "enterprise," "optimism," etc. play little role in the liberal and leftist vocabulary . . . . Art forms that appeal to modern leftist intellectuals tend to focus on sordidness, defeat, and despair . . . . He wrote that leftist's feelings of inferiority run so deep that they cannot tolerate any classification of some things as successful or superior and other things as failed or inferior . . . . There were the theories that soared free of any possible accumulated experience: To spread the revolution, "Revolutionaries should have as many children as they can." There were the procedural afterthoughts that suddenly shattered the orderly argument: "The factories should be destroyed, technical books burned, etc." There were cryptic colloquial drops signaling the writer's loss of patience: The direction society would take "if the industrial-technological system survives the next 40 to 100 years." He felt it would be better to dump the whole stinking system and take the consequences. He wrote cogently about freedom of the press and how it is an important tool for limiting concentration of political power but then opined how the explosion of material made available by advanced technology had reduced press freedom to a nominal freedom and his rationalization for killing people: "Take us (FC) for example . . . . In order to get our message before the public with chance of making a lasting impression, we've had to kill people."[1]

I have included this 'Unabomber' chapter because of its relevance to our rapidly advancing technologies . . .

Theodore Kaczynski had something cogent to say and he got our attention through his bloody insane reign of terror. Killing innocent people can never be justified for a cause. We can only hope that such horrible acts are not repeated in the future and that we as a species will be able to control and acclimate to our new information age.

REFERENCES PROPHETIC OR INSANE?

1. Didion, Joan. (April 23, 1998). "Varieties of Madness The Unabomber Manifesto FC." *The New York Review of Books.*

GOVERNANCE AND DEMOCRACY

Since antiquity to the current day nations have always relied on information management. The ability to impose taxes, promulgate laws, take population censuses, and raise an army lies at the heart of statehood. In the twenty-first century democratic openness means more than voting at regularly scheduled times in fair elections . . . the citizens expect to have access to government data and information.

The state has always been the biggest generator, collector and user of information recording births, deaths, marriages, economic figures, licensing, laws and the weather. Without machines the data heretofore has been rather inaccessible because it was difficult to find and printing such vast amounts of aggregated information was difficult at best. Today citizens and non-governmental organizations worldwide are anxious to access public data at the national, state and local level sometimes with the support of elected and bureaucratic officials. "Government information is a form of infrastructure, no less important to our modern life than our roads, electrical grid or water systems," says Carl Malamud, the boss of a group called Public.Resource.Org that puts government data online. He was responsible for making the databases of America's Securities and Exchange Commission available on the web in the early 1990s.

America leads other countries on data access. On his first full day in office Barack Obama issued a presidential memorandum ordering the heads of federal agencies to make available as much information as possible, urging them to act "with a clear presumption: in the face of doubt, openness prevails." This was all the more remarkable since the previous Bush administration had explicitly instructed agencies to do the opposite.

It is now possible to obtain figures on job-related deaths that name employers, and to get annual data on migration free. Some information that was previously available but hard to get at, such as the Federal Register, a record of government notices, now comes in a computer-readable format. It is all on a public website, data.gov. And more information is being released all the time. Within 48 hours of data on flight delays being made public, a website had sprung up to disseminate them. People expect more from government and this cultural change is encouraged by shopping on the Internet and having 'real-time' access to financial information. Cities like San Francisco, New York, Chicago, and Washington DC have openly provided historical crime information, restaurant health compliance scores by location, children's activities, parking space availability, etc. With more eyes viewing and analyzing the data it is assumed that viable solutions on an individual and collective scale will become apparent. This model creates a 'culture of accountability' to some and a 'danger' to others. Privacy and confidential information ranks high on the 'danger' list along with misuse and abuse of delicate information such as pedophiles utilizing the location of children for their own perverted desires or diplomatically and politically as in the Wiki-Leaks revelations.

Most countries lack America's open-government ethos enhanced over many years by laws on ethics in government, transparency rules, and the Freedom of Information Act. Even in America access to some government information is restricted by financial barriers. Court documents and legal records, for example are public and available online from the Administrative Office of the U.S. Courts (AOUSC) but at a cost of eight cents for each page; even the federal government has to pay. Two companies WestLaw and LexisNexis are being paid by the federal government to publish the material online (organized and searchable with the firms' technologies). Those companies, for their part, earn an estimated $2 billion annually from selling American court rulings and extra content such as case reference guides. With such enormous earnings it seems likely that more companies will compete for the revenue thereby lowering the cost.

In "Full Disclosure" it is noted that there have been many other successes in areas as diverse as restaurant sanitation, car safety, nutrition, home loans for minorities and educational performance. Transparency cannot do the job alone since there needs to be a collective to utilize and understand the importance of the information. Business as well as government needs incentives to supply the data as well as penalties for withholding information. The public information released must be used effectively or misinformation and corruption could be the undesired result. Author Archon Fung warns that as governments release more and more information about the things they do, the data will be used to show the public sector's shortcomings rather than to highlight its achievements. Another concern is that the accuracy and quality of the data will be found wanting (which is a problem for business as well as for the public sector). There is also a debate over whether governments should merely supply the raw data or get involved in processing and displaying it also. The concern is that they might manipulate it – but then so might anyone else.[1]

Catastrophes and Threats

Richard Posner in his book "Catastrophe and Response" wrote about the end of our planet by any number of disasters: Collision with an asteroid that could shatter the earth into a thousand pieces. Precipitate global warming that could, paradoxically, turn it into a giant snowball. A runaway particle experiment that could squeeze the planet down to an uninhabitable hyper-dense marble. Other man-made disasters include: a gene-spliced pandemic, a nuclear-winter war, run-amok robots, and self-assembling 'nanomachines,' billionths of a meter across, gobbling up everything in their path until they have consumed all of life. He sees a cloud of extinction events hovering over the planet waiting to happen unless we manage our technology and rethink our priorities on how we live. He sees no guarantees even if we do come to our senses on the misuse/abuse of our ever increasing technologies.

Posner argues that if we or our descendents have a chance to avoid a final annihilation we must come out of our complacency absorbed in each of our daily pedestrian existences as short as they may be. We must also take seriously the discussion of remote possibilities and the real prospect of cataclysms that will impact all of us and not just some faraway-violent-weather-racked country on the other side of the world. The main problem, over and above their mind-bending dimensions, is that these various sorts of mega-catastrophes seem to most people to be far off, unlikely, and/or conceptually out-of-sight. They seem to be beyond a practical response. We are not equipped emotionally or intellectually to think systematically about extreme events.

The dangers of catastrophe are growing for two reasons: 1. the rise of apocalyptic terrorism; 2. the breakneck pace of scientific and technological advance. The cost of dangerous technologies and the skill to employ them are falling (e.g., nuclear and biological warfare) which is placing more of the technologies within reach of small nations, terrorist gangs, and even individual psychopaths. Posner argues that we are paying less attention to potential extinction than we are paying to social issues of far less intrinsic significance although he does imply that these social issues are trivial: race relations; homosexual marriage; the size of the federal deficit; drug addiction; and child pornography.

In distinguishing the threats it becomes obvious that solutions are at the very least, very difficult to envision and implement unless technology is a major contributing factor: tsunamis, earthquakes, volcanic eruptions, glaciations, asteroid collisions, germ-war pandemic, a laboratory accident, genetically modified crops, artificial life, mechanical super-intelligence, species loss, Greenhouse pollution, cyber-terrorism. Regarding a germ-war pandemic he fears "the possibility that science, bypassing evolution, will enable monkeypox to be 'juiced up' through gene splicing into a far more lethal pathogen than smallpox ever was." He gives an example of a laboratory accident in which a shower of quarks in a particle accelerator self-reassembled into "a very compressed object called a 'strangelet' that would keep growing until all matter was converted to strange matter." He posits a similarly generated accident called a 'phase transition' that would rip the fabric of space itself thus destroying all the atoms in the universe. Posner reviews all the threats in a farrago; i.e., piles of facts, speculative calculations, arguments, and policy dicta. The result is interesting but clearly without real useful solutions.[2]

Posner's purpose is to alter attitudes, redirect mind-sets, refocus worries; transform the currents of popular feeling. He does not believe that the modern way of life globally is sustainable as it currently stands. Unfortunately the identified threats and their costs of impact are difficult to assess although Posner does try albeit unconvincingly. He admits the difficulty in determining what proportions of resources should be dedicated to remote disasters that may or may not happen and where this or that catastrophe should rank on a scale of worries.[3]

Scientists, whose "goal is knowledge, not safety . . . cannot be entrusted with the defense of the nation and the human race." The Large-aperture Synoptic Survey Telescope would as we know be an ideal tool for identifying potentially hazardous near-earth objects. The principal advocates of the project, however, are interested not in near-earth objects, but in remote galaxies. Posner feels scientists need to be brought to a more responsible awareness of their social duty – perhaps by a science court manned by "scientifically literate lawyers," perhaps by a federally funded "Center for Catastrophic-Risk Assessment and Response." "Johnny-one-note civil libertarians uttering fallacious slogans," peddling "bromides about free speech," and obsessing over "coercive interrogation" may object that such measures break constitutional norms. But since September 11, "the marginal cost of civil liberties [has] increased dramatically." As the risk is great, so must be the response. In wartime we tolerate all sorts of curtailments of our normal liberties: conscription, censorship, disinformation, intrusive surveillance, or suspension of *habeas corpus*. A lawyer might say that this is because war is a legal status that authorizes such curtailments. But to a realist it is not war as such, but danger to the unusual degree associated with war, that justifies the curtailments. The headlong rush of science and technology has brought us to the point at which a handful of terrorists may be more dangerous than an enemy nation . . . . It has been commonplace since Thomas Hobbes wrote "Leviathan" that trading independence for security can be a profitable swap . . . . Only the will is wanting.[3]

In the world's 'war on terror' many people feel that the real victim is privacy. Watchdog organizations and advocacy groups are concerned that the threat of terrorism will be an excuse to install new surveillance technology, both in the real world and on the Internet, that would otherwise have aroused fierce opposition. 'Smart' closed-circuit television (CCTV) systems have been installed in public places: e.g., it has been used to scan the thousands of faces of people going to watch Super Bowl games as they pass through the turnstiles. The faces are covertly compared with a database of known criminals, using a facial-recognition system. Other installations have scanned faces in other public places denounced by opponents as "digital frisking." In Europe CCTV is far more prevalent in city centers, shopping centers, sports stadiums, and airports justified as anti-terrorist measures. Systems have been installed in London to scan the license plates of all cars entering and leaving the city without any public protest.[4]

Since the 9-11-01 attacks in America Americans have become more accepting of surveillance technology which is deeply troubling to privacy advocates who are concerned about 'mission creep' whereby the database of suspects is broadened to include more common offenders such as petty thieves, wayward spouses, and public figures. Police have been known to use technology for their own use without public knowledge such as using "Easy-pass" on toll roads to calculate a user's speed and mail him/her a speeding ticket when in excess of the posted speed limits. The extended scope of Internet and telephone wiretapping has become a standard government operation not only in capturing potential terrorist before they act but in tracking down and arresting criminals such as pedophiles based on their Internet activity and site visitations. Public opinion does not rush to defend these potential terrorists and potential criminals but what other unknown/unrevealed uses is the wiretapping being used for? Benjamin Franklin once said: "They that can give up essential liberty to obtain a little temporary safety deserve neither liberty nor safety." Surveillance has become much more prevalent as America has accepted a new balance between security and privacy.

Access and Local Pride

In 2004 Gavin Newsom, the young mayor of San Francisco said: "We will not stop until every San Franciscan has access to free wireless-internet service." By doing this he joined a nationwide movement of cities across the country that have planned and initiated the provision of wireless-broadband access for government workers, residents and businesses. The movement has narrowed the so called "digital divide" bringing America more in line with other parts of the developed world. This movement has cut communication costs, improved efficiency of the workforce, and has made possible new services such as allowing parking meters to accept debit and credit cards. Citywide networks was started by Mayor John Street in Philadelphia in 2004 as transmitters were strung across the entire city providing access both indoors and out. Slow and inefficient dial-up

Internet connections became dinosaurs as the concept of free and cheap broadband quickly spread across the entire country.

Although municipal broadband has not worked for every city the mesh networking it utilizes allows large areas to be blanketed with wireless coverage quickly and inexpensively. As its name suggests, a mesh network consists of an array of wireless access points, only a few of which are actually connected back to the Internet via high-speed links (known as "backhaul" connections). The trick is that all of the access points double as relays, passing packets of data to and from their neighbors. Mesh networks can provide coverage in areas, such as sprawling suburbs, where fast copper or fiber-optic connections are hard to come by. They are reliable, since the failure of one or more access points does not bring down the whole network, and they can also route data around obstacles, such as large buildings, which might otherwise block coverage.[5]

During the French-German ministerial conference in April 2005 Jacques Chirac, President of France said: "We must take the offensive and muster a massive effort." He warned about the dangers of losing the battle for the "the power of tomorrow." He and Gerhard Schröder, then Chancellor of Germany endorsed the plan to build a Franco-German Internet-search engine to be called Quaero (Latin for "I seek"). The plan was in response to the technological challenges posed by America, Japan, China, India, and Brazil. In later months following the conference Mr. Chirac said: "We must take up the global challenge of the American giants Yahoo and Google; Culture is not merchandise and cannot be left to blind market forces; We must staunchly defend the world's cultural diversity against the looming threat of uniformity; Our power is at stake." The French were the main financier and developer of Quaero and created the Agency for Industrial Innovation (AII) based in Paris to oversee the project with an initial endowment of $2 billion.[6]

Quaero became a European research and development program with the goal of developing multimedia and multilingual indexing and management tools for professional and general public applications (mobile environments, professional solutions for production, post production, management and distribution of multimedia documents, facilitation of access to cultural heritage such as audiovisual archives and digital libraries). It is a French project with several German partners, private companies, and public research institutes. The search engine application is often cited as a European competitor to Google as well as Yahoo, Bing, and Ask.com. Quaero is meant for multimedia search and is not intended to be a text-based search engine. Operating in several languages it uses techniques for recognizing, transcribing, indexing, and automatic translation of audiovisual documents. There is also automatic recognition and indexing of images.

Image recognition allows users to search using a 'query image' (not just a group of keywords) that in a process called 'image mining' will recognize shapes and colors from the 'query image' to locate and retrieve similar still images and video clips. Using a technique called 'keyword propagation' whenever the software finds an image without a description which contains elements of or completely matches a labeled image, it will append the description from the labeled image to the unlabelled one. 'Keyword propagation' allows for faster searches, enrichment of the web, and linguistic enrichment since the primary interface and query terms were supposed to be in French and German. The project is also working on 'query sound clips' in addition to the 'query image' search with the intention of transcribing their content to text and translating it to other languages. There has been criticism that President Chirac is more interested in defending French pride than in global advancement of the Internet. It has been reported that Quaero has been scrapped as Germany will be producing its own 'scaled down search engine.' [6]

e-Government

Worldwide governments have spent a lot of money putting government services online with results that are difficult to measure and often disappointing. Accenture has pioneered the business of selling technology services to government, publishing reports on the subject since 2000. By 2004 the term "e-government" was becoming stale. Greg Parston of Accenture says the company now places "much greater emphasis on outcomes for citizens." Many other companies are publishing similar reports – and trying to make the most of public-sector technology contracts. Not-for-profit bodies such as the United Nations, the World Bank, many universities and think-tanks use similar jargon and make similar points on the subject.

Accenture's 2007 report, for example, highlights four "pillars of leadership in customer service:" a "citizen-centered perspective," a "cohesive multi-channel service," "fluid cross-government service," and "proactive communication." Bureaucracy like well-run businesses can take advantage of the technology to not see customer complaints as a nuisance but to use the data gathered from websites and call-centers to fine-tune the products and services that they offer. A good e-government scheme starts off from the citizen's eye view, not the bureaucrat's one. Similarly, a "cohesive multi-channel service" simply means that whether you make contact by Internet, telephone, letter, or personal visit, you should get the same efficient service. The private sector has been doing this for a long time: you can shop by looking at a catalogue or browsing in a bricks-and-mortar store, then order online, phone with a query, receive your goods by post and take them back to a shop if there is a problem. That sort of integration for governments is still difficult to achieve.

In principle, there is not much argument about the desirability of putting government online. Technology helps to make public administration more open, more responsive and cleaner. Taxpayers save money. Citizens get better services. Democracy is revived. First world countries already have the broadband penetration, computer literacy, and skilled bureaucrats needed for sophisticated e-government. Third world countries may have more to gain via transparent online government by possibly removing their perceived greatest barrier to development; wasteful, incompetent, and corrupt public administration. People cannot steal public funds for too long when some system is put in place to control and disseminate the information about the theft. Shame is a powerful deterrent and can overcome greed.

Unlike e-commerce which has been a spectacular success e-government has not transformed public administration although it has spent huge amounts of taxpayers' money on big computer systems poorly thought out and overpriced. Rarely do the promised benefits materialize. For example in seven years Britain wasted £2 billion on projects that have ended up being cancelled and written off.

The 'digital have-nots' in this digitally technological advanced age have dropped further in comparison to the articulate and well-connected (literally) part of the population thus accentuating unfairness and less openness. Putting public services online is no use to those who cannot afford a computer or will have nothing to do with technology. The phrase "look on our website" is a turn-off for a significant chunk of most countries' population. Where the Internet is used to increase public participation in democracy, the problem is sharper still for those at the margins who go unheard.

Technology may have given citizens a bit more information about government, but it has given government a lot more information about them, for good or ill. Efficient government can be repressive government. In Germany, and many other countries, the law says that people must tell the government where they live. In America, that would be an outrage. Face-recognition software, remotely readable chips in passports, recognition technology for car license-plates and biometric identity checks offer endless possibilities for controlling the citizen. George Orwell's "1984" is an example of malign e-government: a screen in every dwelling monitors the inhabitants' doings with an efficiency that would thrill today's operators of Closed Circuit TV systems. In First World countries taking out lots of low-paid, low-skilled jobs in government adds to the problems of a section of the workforce that already resents the march of progress. Just as technological change has stoked protectionist sentiment in the private sector modernization of government can seem threatening and unfair to those who work in the public sector.

The biggest concern about bad e-government is that it is worse than useless. When humans make mistakes, you can argue with them. When a computer insists that you owe the government money, your car is illegally parked, or you do not exist, unscrambling the problem is much harder. Similarly, when records were kept on paper by real people, the scope for error was limited. At worst, a pile of confidential information might be thrown away by accident and end up on a rubbish dump. Data might be wrongly entered or incorrectly filed and never found again. That was unpleasant for the individuals concerned, but usually nothing worse. Digital mistakes can be much more serious. In November 2007, for example, the British government managed to lose two discs of data containing the (unencrypted) personal and financial details of 25 million households. Nobody has been able to explain why the data – which were being sent from one part of the public sector to another – were not transmitted using the secure government intranet, a computer network specially designed for this sort of job and established at the cost of tens of millions of pounds. In the hands of fraudsters and identity thieves, the sort of data that a government can collect by law could be misused with disastrous results.

Those most closely involved in e-government are aware of the problems. As a 2007 Accenture report points out, they have become extremely cautious about what they can deliver in the short term. "After the splash of creating exciting visions and promises of truly citizen-centric government services, many governments now . . . find themselves . . . playing catch-up on a promise that citizens expect to be fulfilled . . . The essential infrastructure work that comes next is unlikely to capture the imagination of citizens and the media. It is hard work, plain and simple."

The world of e-government moved forward with the Organization for Economic Co-operation and Development's (OECD) first publication of "Government at a Glance" in 2009. Published annually with a range of international comparisons the OECD's sharpest analysts and statisticians are now trying to crack a question which seems blindingly obvious yet which hardly anybody to date has tried to answer properly: what are the costs and benefits of e-government? Australia is one of a small handful of countries (which also includes Britain) that have tried to assess the aggregate effects of e-government. The answers were mixed. Australia's National Office for the Information Economy in 2002 found that the majority of 38 e-government projects surveyed were likely to make things cheaper and more convenient for officials, users or both. But for the 24 projects that were expected to result in specific cost reductions or higher revenues, the total investment of $108 million produced a saving for the government of only $100 million. It does not suggest that e-government is an easy way for the state to balance its books.

The most obvious benefit of e-government is to reduce the burden for citizens, taxpayers and so on of dealing with government by gaining from information online. However, when technology not just distributes information to the public but collects and uses it as well, the issues become much more complex. Where data-sharing laws permit, individuals and businesses may save money and effort by having to provide information about themselves to the authorities only once rather than many times over. And on the government side, dealing with people via automated online systems is much cheaper than face to face or on the phone. Unlike the private sector that can insist that all customers use the Internet, governments have to keep all channels of communication open. In most countries the heaviest consumers of public services, the old and the poor, are the least likely to use the Internet. The costs of providing both on and off line services may not save the cost of implementing the new technology.

The non-financial benefits of e-government is simply demonstrating to the public that government works cleanly, quickly, and efficiently could pay off in votes for the politicians, especially those who have pushed through the necessary reforms. It is hoped that it will increase trust and reduce perceptions of corruption. Such measures play an important part in countries' global competitiveness rankings, and a good performance in this area makes a country more attractive to foreign investors.[7]

Every nation since the formation of nation-states in antiquity has always been a product of information management. The ability to impose taxes, promulgate laws, count citizens, and raise an army lies at the heart of statehood. Today democratic openness encompasses more than voting in free and fair elections at regular intervals; citizens expect access to government data. The state as the biggest generator, collector and user of data keeps records on every birth, marriage and death, compiles figures on all aspects of the economy, keeps statistics on licenses, laws and the weather. It is only with the new digital technology that the notoriously difficult task of finding and aggregating lots of printed information is rapidly becoming less appealing. Citizens and non-governmental organizations the world over are pressing to get access to public data at the national, state and municipal level. America is in the lead on data access.

"The city is facing its eighth budget shortfall. We're looking at a 50% reduction in operating funds," says Chris Vein, San Francisco's CIO. "We must figure out how we change our operations." He insists that providing more information can make government more efficient. California's generous "sunshine laws" provide the necessary legal backing. Among the first users of the newly available data was a site called "San Francisco Crimespotting" by Stamen Design that layers historical crime figures on top of map information. It allows users to play around with the data and spot hidden trends. People now often come to public meetings armed with crime maps to demand police patrols in their particular area. The point of open information is not merely to expose the world but to change it. In recent years moves towards more transparency in government have become one of the most vibrant and promising areas of public policy. Sometimes information disclosure can achieve policy aims more effectively and at far lower cost than traditional regulation.

In an important shift, new transparency requirements are now being used by government – and by the public – to hold the private sector to account. For example, it had proved extremely difficult to persuade American businesses to cut down on the use of harmful chemicals and their release into the environment. An add-on to a 1986 law required firms simply to disclose what they release, including "by computer telecommunications." Even to supporters it seemed like a fudge, but it turned out to be a resounding success. By 2000 American businesses had reduced their emissions of the chemicals covered under the law by 40%, and over time the rules were actually tightened. Public scrutiny achieved what legislation could not. Public access to government figures is thought to release economic value and encourage entrepreneurship as has already happened with weather data and with America's GPS satellite-navigation system that was opened for full commercial use in the late 1990's. Many firms are making a good living out of searching for or repackaging patent filings.

John Stuart Mill in 1861 called for "the widest participation in the details of judicial and administrative business . . . above all by the utmost possible publicity." These days, that includes the greatest possible disclosure of data by electronic means.[8]

Privacy, Security, Processing, Smart Grids

Gutenberg invented movable type in the mid-1400s; there were plenty of books around, but they were expensive and poorly made. The first copyright law in the early 1700s in

Britain was designed to free knowledge by putting books in the public domain after a short period of exclusivity, around 14 years. Laws protecting free speech did not emerge until the late 18th century. Before print became widespread the need was limited.

Today with astronomical information flows the relationship between technology and the role of the state is changing once again. Privacy laws were not designed for networks. Rules for document retention presume paper records. Since all the world's information is interconnected, it needs global rules. New principles for an age of big data sets will need to cover six broad areas: privacy, security, retention, processing, ownership, and the integrity of information.

The major area of concern is privacy since people are disclosing more personal information than ever. Social-networking sites and others actually depend on this disclosure. As databases grow, this information can often be unlocked with a little computer effort. The tension between individuals' interest in protecting their privacy and companies' interest in exploiting personal information could be resolved by giving people more control. They could be given the right to see and correct the information about them that an organization holds, and to be told how it was used and with whom it was shared.

Better technology should eliminate privacy concerns since firms are already spending a great deal on collecting, sharing and processing the data; they could divert a sliver of that money to provide greater individual control. The benefits of information security – protecting computer systems and networks – are inherently invisible: if threats have been averted, things work as normal. That means it often gets neglected. One way to deal with that is to disclose more information. A pioneering law in California in 2003 required companies to notify people if a security breach had compromised their personal information, which pushed companies to invest more in prevention. The model has been adopted in other states and could be used more widely.

In addition, regulators could require large companies to undergo an annual information-security audit by an accredited third party, similar to financial audits for listed companies. Information about vulnerabilities would be kept confidential, but it could be used by firms to improve their practices and handed to regulators if problems arose. It could even be a requirement for insurance coverage, allowing a market for information security to emerge. Current rules on digital records state that data should never be stored for longer than necessary because they might be misused or inadvertently released. But Viktor Mayer-Schönberger of the National University of Singapore worries that the increasing power and decreasing price of computers will make it too easy to hold on to everything. In his recent book "Delete" he argues in favor of technical systems that 'forget' digital files that have expiry dates or slowly degrade over time. Yet regulation is pushing in the opposite direction. There is a social and political expectation that records will be kept. Peter Allen of CSC, a technology provider believes the more we know, the more we are expected to know . . . forever. American security officials have pressed companies to keep records because they may hold clues after a terrorist incident. In the future it is more likely that companies will be required to retain all digital files, and ensure their accuracy, than to delete them.

Processing information is another concern. Some are concerned that algorithms will replace human intuition and the legal implications of using statistical correlations. There is the worry about the "ethics of super-crunching." Examples abound such as racial discrimination in American loan applications coming from a highly correlated variable such as the education level of one's mother. If computers can predict an individual's susceptibility to a disease from other bits of information it should not be difficult to predict a person's predisposition to committing a crime. Regulating this by disallowing discrimination on the basis of something that might or might not happen will be the way we will go but regulation does not always insure compliance. Illegal immigrants have been deported after taking part in census surveys although regulation protected them from such acts if they participated in providing the information. Privacy rules lean towards treating personal information as a property right but this right has considerable problems with compliance and enforcement. The philosophy of democracy and freedom are noble concepts but ensuring the integrity of information is far from insured. Secretary of State Hillary Clinton attacked the Chinese when they hacked into Google's computers but their obsession with censorship pollutes the Internet environment without regret or guilt. International co-operation has historically proven to be a difficult concept at best.[9]

McAfee, a popular computer security and anti-virus firm reported that China has conducted cyber-espionage through a hacking campaign labeled "Night Dragon." The purpose of the campaign is to somehow give China's energy companies a competitive edge by stealing bid documents and other sensitive information from global oil-and-gas companies. China ensures all its country's energy majors have a low cost of capital whenever they bid for global assets and knowing what the other bids are gives the Chinese companies a seemingly insurmountable advantage. This use of technology by an autocratic rising world power stokes a general sense of unease throughout the world. In 2010 Chinese companies represented twelve percent of the oil-and-gas global asset market.[10]

Utilities worldwide without technology have and are still causing significant problems as pipes and cables burst or decay with age causing emergency calamities. Thames Water in London replaced over 1,300 miles of cast-iron Victorian mains (those most likely to break) with plastic ones equipped with wireless sensors giving it a better view of the entire network. By knowing what is going on in its network the company is able to act quickly on the information and prevent emergencies from growing into disasters. Infrastructures throughout the world need this kind of technological information in order to be more efficient and prevent loss of property and life. Whether in water, power, transport, or buildings, all are trying to turn their dumb infrastructures into something more like a central nervous system. This activity is the convergence of the physical and the digital world. Putting sensors and actuators (devices to control a mechanism) into physical infrastructures is not exactly new. Known as "supervisory control and data acquisition," such systems have been around for decades.

But many still require human intervention: workers have to be sent out to download sensor readings or to fix problems. And even if sensors and actuators are connected, different types often feed into incompatible systems, so they cannot be easily combined to automate processes.

The ideal is remote central control to monitor and even fix remotely. Thames Water is investing millions to take action remotely and automate a lot of its processes. If the project works, the system will not only automatically deal with leaks but also schedule work crews and send text messages to affected customers. One day soon Thames Water may even be able to send out work crews before a main actually breaks. In early 2010 the firm began using a web-based service provided by TaKaDu, an Israeli company, that acts as the network's "eyes and ears." The firm analyses historical and online data to provide a basis for comparison, enabling its algorithms to detect things that are about to go wrong.

For infrastructures to become truly smart, however, it is not enough to put more intelligence into the core of a network. The edge – the interface with users and devices – also has to become clever. This is the idea behind smart metering, which has made a good deal of progress in the power industry. Smart meters and other gear needed to make grids more intelligent will not come cheap. Morgan Stanley, an investment bank, predicts that the worldwide smart-grid market alone will grow from $20 billion in 2010 to $100 billion in 2030. Yet the benefits also promise to be huge: power savings, reduced investment in electricity generation and lower carbon emissions. A mechanism called "demand response" tells systems when to shut themselves off when demand for electricity is high as during peak periods of utilization. Telling customers to turn off systems does not insure compliance. The ultimate point of smart grids, however, is to allow dynamic pricing, with electricity charges fluctuating in response to demand. This could cut power demand during peak hours. The main objective of smart power meters is to lower the peak load and thus enable utilities to keep down their peak generating capacity.

In transport the equivalent of a smart meter is a vehicle's on-board unit that can pay tolls or keep track of a vehicle's location through the satellite-based global positioning system (GPS). Paying tolls has proliferated. The city-state Singapore not only charges drivers for using much-travelled roads it also adjusts traffic lights to suit the flow of vehicles, uses data collected by taxis to measure average speed, and is developing a parking-guidance system, noting that cars looking for somewhere to park are now a big cause of congestion. It is working on introducing real-time dynamic pricing of toll roads that changes according to how congested a road is at any given time. Singapore may also become a model for the world with water: waste water is made potable after extensive treatment creating a closed loop system where 'every drop counts.' By inaugurating a tidal barrier in 2008 Singapore keeps seawater out of its harbor thus creating a sizable reservoir to catch, use, and reuse rainwater. The city-state's desalination plants are also among the world's most efficient. All this means that the island – smaller than Luxembourg and home to nearly 5 million people as well as an economy nearly as big as that of Hong Kong – is able to meet more than 60% of its water needs on its own. But it wants to go even further: 50 years from now it hopes to be self-sufficient.[11]

Talking about water in Singapore quickly turns political since it hopes to become self sufficient when its water-supply agreement with Malaysia expires in 2061. More than once Malaysia has threatened to increase prices or even cut off supplies. Singapore either gets too much water and floods or too little water causing a draught so politics is not its only obsession about water; the issue of scarcity and excess play large. The city wants to become a 'living laboratory' for smart urban technologies of all kinds – not just water and transport systems but green buildings, clean energy and city management.

There is strong demand for making cities smarter, not just in China and other rapidly urbanizing countries but throughout the Western world. 'Smart-city' projects have been multiplying around the world and have one thing in common: they aim to integrate the recent efforts to introduce smart features in a variety of sectors and use this 'system of systems,' as IBM calls it, to manage the urban environment better. The best-known smart city is Masdar, a brand-new development in Abu Dhabi that hopes to eventually become home to 40,000 people. It is being built entirely on a raised platform, which makes maintenance and the installation of new gear much easier. Below the platform sits the smart infrastructure, including water pipes with sensors and a fiber-optic network. Above it is to be a showcase for all kinds of green technology: energy-efficient buildings, small pods that will zoom around on paths (no cars will be allowed) and systems that catch dew as well as rainwater.[12]

REFERENCES GOVERNANCE AND DEMOCRACY

1. Fung, Archon, Graham, Mary, Weil, David. (2007). *Full Disclosure The Perils and Promise of Transparency*. Boston, MA: Cambridge University Press.
2. Posner, A. Richard. (2004). *Catastrophe: Risk and Response*. USA: Oxford University Press.
3. Geertz, Clifford. (March 24, 2005). "Very Bad News" *The New York Review of Books*.
4. *The Economist* (September 20, 2001). "Surveillance technology . . . Uncle Sam and the Watching Eye."
5. *The Economist* (March 9, 2006). "Wi-Pie in the sky? Communications: Cities across America plan to build municipal Wi-Fi networks to widen access to broadband. Will they work?"
6. *The Economist* (March 9, 2006). "Attack of the Eurogoogle . . . Search technology: Can an ambitious new European search engine, backed by the governments of France and Germany, challenge Google?"
7. *The Economist* (February 14, 2008). "A survey of technology and government . . . The good, the bad and the inevitable . . . The pros and cons of e-government."
8. *The Economist* (February 25, 2010). "The open society . . . Governments are letting in the light."
9. *The Economist* (February 25, 2010). "New rules for big data . . . Regulators are having to rethink their brief."
10. Denning, Liam. (February 10, 2011). "Hacking Into the Dragon's Energy Advantage" *Wall Street Journal*.

11. *The Economist* (November 4, 2010). "Making every drop count . . . Utilities are getting wise to smart meters and grids."
12. *The Economist* (November 4, 2010). "A special report on smart systems . . . Living on a platform . . . For cities to become truly smart, everything must be connected."

TECHNOLOGY

In his seminal book "Mirror Worlds" David Gelernter, professor of computer science at Yale University and survivor of a Unabomber attack predicated that we "will look into a computer screen and see reality." "Some part of your world – the town you live in, the company you work for, your school system, the city hospital – will hang there in a sharp color image, abstract but recognizable, moving subtly in a thousand places."[1]

Similar but not exactly like the science fiction movie "The Matrix" Dr. Gelernter's writing has become more and more accurate. We all know our real world that we live, play, and work in and are becoming more familiar with its digital reflection that takes real information and automatically acts on it. Our real world has become inundated with sensors, picking up everything from movement to smell. If a door opens in the real world, so does its virtual equivalent. If the temperature in a house falls below a preset level the mirror digital world automatically turns on the heat until the real world is heated to a predetermined temperature. Humanity is building more and more "mirror worlds," or "smart systems," as we now call them. Due to the ubiquitous rise of technology in all areas of human endeavor the real and digital worlds are converging via connected sensors, cameras, wireless networks, communications standards, and all our various activities.

The convergence of these two worlds (real and digital) is not fully recognized by most people because it is happening everywhere and is not fully understood for what it is. Controlled environments are utilizing it extensively. The large conglomerate Siemens maintains virtual replicas of factories to monitor and change their configurations as necessary. It is spreading everywhere and is developing its own language: 'virtualization of the real world;' 'cross reality;' 'augmented reality;' 'alternate reality.' Google's Earth and Street View services are the first, if static, replicas of the entire world. Sensors placed in cows allow the tracking of every movement from birth to abattoir. Smart power meters tell utilities how much electricity is used in real time.

Smart Systems

As I write the real culprit in speeding up the convergence of the real and digital world is the ubiquitous smartphones and their applications (aka 'apps') that can easily be downloaded from the Internet. Smartphones abound with sensors that measure many things including brightness and geographical location. The 'apps' in smartphones are miniatures of larger smart systems that control appliances and give precise data on geographic locations.

Smartphones now provide 'augmented reality' which through a video camera in the phone and a downloaded app the phone owner points at a street, and the software will overlay the picture on the screen with various pieces of digital information (e.g., names of streets, businesses, houses for sale, etc.). Since Moore's law holds that processing power doubles every eighteen months and applies to all types of electronic equipment including sensors this type of 'societal information-technology systems' (SIS) is bound to grow very rapidly. Exponentially increasing processing power and improved connectivity allows computing systems to store and analyze huge amounts of data from sensors and other devices. Networks of data centers packed with thousands of servers known as 'Computing clouds' are being put together worldwide. These networks store and analyze data to allow smart systems to react instantly to any environmental change.

Information-technology (IT) firms are working fast and furious with smart systems to increase the intelligence of our energy use, transportation, and cities. Some companies are initiating plans to deploy their ever increasing knowledge of certain industries such as health care and manufacturing. Many countries have been spending large chunks of their stimulus packages on smart-infrastructure projects, and some have made smart systems a priority of industrial policy. In many countries the physical infrastructure is ageing, health-care costs are exploding, and money is tight. Using resources more intelligently can greatly assist in a time of economic slowdown. Monitoring patients remotely can be much cheaper and safer than keeping them in the hospital. A bridge equipped with the right sensors can tell engineers when it needs to be serviced. Technological advances have gained momentum because there is a real need for more efficient systems.

China is a good example of the need for more efficient systems since it is becoming urbanized on an unprecedented scale. By 2025 it is estimated that more than the current population of the United States will have moved to its cities. These numbers will require incredible efficiency to provide enough food, transport, electricity, and water for everyone. It is argued that smart systems may well be humankind's best hope for dealing with its pressing environmental problems, notably global warming. Most power grids today are essentially 'dumb,' i.e., lacking in any intelligence for our transport and water-distribution systems. It is seen by many to be a 'no brainer' when it comes to utilizing technology for what ails humanity. What is there to lose? Privacy and the risk of abuse by corrupt governments usually are first on the list. The smart systems today make the ubiquitous tele-screen monitoring device in George Orwell's novel "1984" a novel plaything. Smart systems could gang up on their creators, as they did in "The Matrix," the 1999 film in which human beings are plugged into machines that simulate reality to control humans and harvest their bodies' heat and electrical activity. No one is forecasting "The Matrix" scenario but smart systems are vulnerable to malfunctioning and hacker attacks.

Another concern is the inequality posed by smart systems in that people with access will be enormously better informed than those without access giving them a very significant advantage in life and greatly broadening an already vastly unfair and unequal world.

The concerns are many and unless they are dealt with the world may experience a significant neo-Luddite reaction.[2]

Sensors

Anything and anyone – machines, devices, everyday things and particularly humans – can become a sensor, gathering and transmitting information about the real world. Radio-frequency identification (RFID) are electronic labels being used to identify everything from cattle to tombstones. They are very difficult to recycle and soon may have to be separated in our garbage along with glass, metal, and paper. RFID tags are reliable, inexpensive and do not require a power source (when exposed to a radio signal they use its energy to send back information . . . usually a long number identifying an object). They were created in the late 1980's in planning for a future in which the real and virtual worlds would be connected. RFID's have evolved into 'active' tags (which have their own power source) and wireless sensors which are evolving from the small to the tiny. These devices can sense everything from rare chemicals to unusual bacteria. Biosensors can now identify thousands of viruses and bacteria.

The development of sensors will be increasing exponentially as they start to power themselves by scavenging energy in their environment (e.g., in the form of light and motion). Radio spectrum is another scarce resource that is being handled by smart power meters that relay their readings via 'mesh networks.' Engineers are forecasting the development of 'smart dust' in which sensors as small as dust particles can be dispersed to monitor people or troop movements. Silicon Valley now has devices one thousand times more sensitive than those in our smartphones with the intent to cover the world with over one trillion sensors that will deliver data to anybody who needs it (e.g., carmakers, military, governments, retailers, prospectors, scientists, etc.).

Digital cameras have become the most widely deployed sensor thanks to its inclusion in nearly all mobile phones. Many objects no longer need an electronic tag or even a barcode to be automatically identified. Goggles, a service offered by Google, can recognize things like book covers, landmarks, and paintings simply by having users send the picture instead of typing in search words. Many of today's machines and devices already have data-generating digital technology and are becoming more and more commonplace. Body scanners in hospitals, aircraft engines, refrigerators, and coffee machines are a few examples of machines that can 'phone home' and provide information.[3] Smartphones are turning humans into the best sensors since they provide information without any extra effort. Via GPS they provide geographical location data that can provide directions or help avoid traffic jams. This 'crowdsensing' has become very common throughout the world as more and more people are willing to actively gather and upload data. They are also the eyes (cameras, emails, and text messages) and ears (audio recordings) for mountains of information. People have become the sensory organs of the Internet.[4]

Twitter, a micro-blogging service with hundreds of millions of users sends out hundreds of millions tweets a day whenever they see, hear, or read something. Twitter has allowed ordinary citizens to hear and report breaking news stories before the mainstream media sources. Other companies and websites record people's wanderings and locations and apps have been developed to report safety issues in the environment or things that need repair or attention (streetlights, or fallen trees).

Stream Computing and Data Visualization

Tools are being developed to deal with this data deluge through 'stream computing' that can ingest thousands of 'data streams' and analyze them almost instantaneously. Natal care units are testing such systems to monitor babies born prematurely. They take in streams of biomedical data, such as heart rate and respiration, and alert doctors if the babies' condition worsens. Analytical software that crunches 'structured' data has been around for many years providing information on fraudulent credit cards and predicting the demand for flights during a public holiday. Analytical software can now interpret 'unstructured' data, mainly free-form text as in social networks such as Twitter and Facebook. The sentiment of chatter and influential comments for example allow companies to target individuals for marketing messages. The main goal of smart systems is 'to close the loop' which means using the knowledge gleaned from data to optimize and automate all kinds of processes. The number of potential applications is vast, ranging from physical infrastructures and manufacturing to preventing car accidents.[3]

With the seemingly infinite amount of data being produced it is necessary to pare down the information so mere humans can somehow digest it. Visualizing data deals with the inhuman scale of information so that the human eye can see it. Data-visualization specialists deal with this challenge by developing display software. They are skilled in computer science, statistics, artistic design, and storytelling. Stock market information for example is vast and hard to display but at least most of it is numerical. Words are more difficult; one way of depicting them is to count them and present them in clusters with the more common ones shown in a larger font called a 'word cloud.' This popular method gives a rough indication of what a body of text is about. In 1998 Martin Wattenberg came up with the idea that launched the field of data-visualization. A graphic designer with the magazine SmartMoney he created a 'map of the market' in the form of a grid that showed more than 500 companies arranged by sector with shades of green and red to indicate a rising or falling price. Wikipedia, the online encyclopedia written by volunteers needed a way to monitor edits. They used another way to present text (devised by Martin Wattenberg and Fernanda Viegas) that creates a permanent record of every edit to show exactly who changed what, and when . . . once again enormous amounts of data. Different colors are assigned to different users to show how much of their contribution remains by the thickness of their representative line. Line Zigzags indicate entries being repeatedly removed and restored over time. The way articles are categorized uses another software to improve categories by showing modifications as a sea of color. Some data-visual works have been exhibited in places such as the Whitney Museum and the Museum of Modern Art in New York. Visualization is a relatively new discipline. If in the old saying 'a picture is worth a thousand words,' than an 'infographic' is worth an awful lot of data points.

Data-visualization is only a natural outcome of processing information since the human brain finds it easier to process information if it is presented as an image rather than as words or numbers. The right hemisphere recognizes shapes and colors. The left side of the brain processes information in an analytical and sequential way and is more active when people read text or look at a spreadsheet. Looking through a numerical table takes a lot of mental effort, but information presented visually can be grasped in a few seconds. The brain identifies patterns, proportions, and relationships to make instant subliminal comparisons. Businesses care about such things and hire psychologists to design their charts and color schemes.

These graphics are most helpful when looking at immense quantities of data. For example one hundred years of census data, every road in America based on the population density, commercial flights over a twenty-four hour period, cost to users and insurers of various diseases throughout a lifetime. Data-visualization software has become more accessible (free and paid-for products) becoming for visualizing data what word-processing did for text. As print publications move to e-readers, animated infographics will eventually become standard. Mapping social interactions can make a difference by enabling people to understand complex matters and find creative solutions, especially in the workplace with many people engaged on a single project.[5]

Apple Computer launched its revolutionary tablet computer the iPad in 2010. Apple had filed a number of patents since 2006 regarding its intentions to move into touch computing that started with the iPhone. The iPad was preceded by Amazon's Kindle and Barnes and Noble's Nook. In 2010 hard book publishing was bigger than both the music and film industries with e-books just beginning to become more popular. In 2009 to most people's surprise book apps (i.e., books that can be read electronically on iPad and iPhone) surpassed the number of game apps in Apple's own App Store. It was assumed that most people were not reading but digital readers seemed to be reading a lot. The avid hard book readers argued that there is something deeply satisfying about a 'real' book, a book made of pages bound between hard and soft covers, into which you slip a bookmark, fan pages, crack and fold the binding as you read through the text. E-books allow you to carry thousands of books in our pocket but do not fill up your bookshelves. If the past predicts the future then with time hard-book readers will go the way of the horse and buggy.

iPad and the iPhone come with a Web browser app that takes the user directly to the Internet making the devices comparable to any computer. You can check email, listen to the radio, watch a movie, play poker, read the headlines, edit photos, compose a song, shop for shoes or clothing, track calories, look up recipes, read a book, write a book, etc. What is revolutionary about iPad is that app development can encompass all other devices into one. Apple has proprietary rights over its products with the clear intent on profit whose app development and corporate deals could compromise access to certain sites. The Internet, on the other hand is an unrestricted playground accessible to just about anyone; it is all about freedom and openness that has sparked creativity and ingenuity. Some governments fear it and some corporations are threatened by it.[6]

Genetic Engineering and Nanotechnology

Technology has had a profound impact on the production of our food through genetic engineering. We have genetically been modifying organisms since the domestication of plants and animals. We have developed radically new ways to manipulate heredity and call it genetically modified organisms (GMO). The ancient modifications produced organisms not very different from their wild ancestors but have many characteristics that may be opposite of the organisms from whence they came. Large kernel maize for example is great for long periods of storage but on its own would soon disappear since it cannot disperse its seeds. The history of domestication is the history of genetic modification of organisms to make them more unnatural. In the past new varieties of plants and animals have come from desirable variants propagated selectively. Variation within species can be augmented by mating with other closely related species that do not normally interbreed; ergo the essence of domestication. The ideal of genetic engineering is to remake an organism so as to produce all the desired variants in the new organism; e.g., disease resistant wheat with high yield wheat to produce disease resistant high yielding wheat. Modern genetic engineering extracts DNA corresponding to a particular gene from a donor organism and inserts it into the cells of another so it becomes part of the recipient's genome. Transgenic organisms result when the source of the DNA is a distant incompatible species; the donor and recipient do not need to be anything like one another. The human gene for insulin is injected into the genome of bacteria which grow in industrial vats and produce human insulin for the market; no one has been harmed from this bacterially produced insulin. Transgenic DNA transfers in agriculture has been utilized to provide crop plants with resistance to insects and herbicides used to control weeds. There is no general rules of what will work since any gene in any species can be transferred to other species. Some of the transfers can be harmful or even lethal to the recipient organism and would have no practical known value. Hundreds of plant varieties created by genetic engineering have been tested under federal regulatory agency guidelines. Several dozen transgenic varieties are commercially available: corn, cotton, squash, potatoes, canola, soybeans, and sugar beets. The federal government utilizes the National Academy of Sciences through its research arm, the National Research Council (NRC) to produce an expert report to guide regulatory policy. Without formal request the NRC has produced several reports since 1974 on genetic engineering that involve agriculture, transgenic varieties, and environmental threats to human health. The costs and benefits of genetic engineering has not been fully explored. Like so much of cost-benefit analysis no really useful or meaningful results have been published.

The contentious issues in GMO's are: 1. threats to human health; 2. possible disruption of natural environments; 3. threats to agricultural production from a more rapid evolution of resistant pests; 4. disruption of third-world agricultural economies; 5. principled objections to 'unnatural' interventions. The NRC reports deliberately exclude numbers four and five and states: "The study does not address philosophical and social issues surrounding the use of genetic engineering in agriculture, food labeling, or international trade in genetically modified plants." The Academy follows a fundamental principle in all of its reports that it is the product and not the process that matters. The NRC is not concerned with whether a product came from conventional manipulations or by transgenic transfer of DNA. It is only concerned whether the new property of the resultant organism is harmful to health and the environment.[7]

The NRC authors point out that the conventional methods of breeding including transgenic transfer might produce varieties with toxicity to humans or other species, or with unusual invasive abilities, or with greater resistance to pests that would hasten the evolution of more bothersome pest species. There are many examples of newly developed troublesome weeds that have come about from the hybridization of crop plants with their wild relatives and several where rare wild species have become extinct by the hybridization of plants. The big concern is the constant spread of millions of pollen grains produced by hundreds of acres of a transgenic crop that will produce hybrids with weedy species at the margins of cultivated fields eventually resulting in a new weedy invasive or competitive form. These examples can be found in Rissler and Mellon's book "The Ecological Risks of Engineered Crops."[8] The only "adverse effects of agricultural varieties on any animal or plant species in nature, including human health, have been from conventionally bred organisms or from the introduction of invasive species from distant geographical areas, or from foods like peanuts and milk which some people are naturally allergic."[7]

"No unequivocal conclusions can be drawn about the overall effect of genetic engineering technologies. It is clear that any manipulation of organisms, whether by conventional means or by genetic engineering, poses some danger to human health, to present systems of agricultural production, and to natural environments. All of these potential effects have led to a fairly effective apparatus of government regulation whose chief deficiency is its dependence on data supplied to it by parties whose prime concern is not the public good but private interest. Nothing is significantly changed in this situation by the introduction of genetic engineering. The technology provides a method for transferring a specific gene into a crop, rather than the uncontrolled mixture of entire genomes that takes place when two varieties or species are crossed. On the other hand the random disruptions of regulatory genes of the recipient that may take place are totally uncontrolled. On balance, it is impossible to say whether we have achieved greater or lesser control over the unintended consequences of mucking around with nature."[7]

Bill McKibben wrote "The End of Nature" about how the human race has been rearranging the biosphere with the aid of genetically modified plants to suit what it believes is in its own interest. In "Enough" he takes on the topics of immortality, eternal youth, godlike beauty, and hyper-intelligence. He denounces the whole enterprise of science that experiments with these goals and prophesizes our doom. He addresses the greed in all of us and why we should stop wanting more and more. He knows that 'more' is an appropriate response to 'not enough' but not the 'more' associated with pure greed and dedicated to endless accumulation and power. As his book title implies 'enough' is already a feast because it is 'us' as we are with a few allowable improvements. Mores are very tempting to cross when it comes to improving our human entities and the scientific landscape is very crowded with them. Human alterations include: genetic alteration or gene splicing; nano-technology; cybernetics; and cryogenics. Genetic alteration can lead parents to select the physical and mental traits they want their children to possess. Nano-technology is the development of single-atom machines that can replicate themselves and assemble and disassemble matter. The film "Fantastic Voyage" miniaturized an entire surgical team to enter a human body and make repairs. Cybernetics is the melding of man with machine as in the "Six Million Dollar Man." Machine hearts have been in use for several decades. Cryogenics is the science of freezing your body or your head until such a time as science has progressed to the point that you can be restored to health. Your head (if that is all you can afford to freeze) can grow the full body again from your or someone else's DNA.

We have genetically engineered food, insects, and small animals. Homo Sapiens cannot be far away. It is in our nature to question, experiment, and make improvements. We have a very high opinion of ourselves ever since our Creator gave us dominion over our planet and everything on it. He told us we were created in His image and ergo many believe we can create life as He did. The dangers and risks are many as dramatized in Mary Shelley's "Frankenstein." We can insert desirable genes into an egg and implant it into a womb or a test tube and create a customized child. McKibben argues that doing such will eliminate the joy and mystery of life because we will not have to strive for anything because it will all be preprogrammed in us before we were born. We will never know if our emotions are truly ours or embedded memories like the 'replicants' in the movie "Blade Runner." Whatever parents want of their progeny will no longer be the wishes in dreams but in the real existence of their children. In our competitive world each new generation of babies (as with our machines . . . computers, smartphones, sensors, etc.) will have to have the latest enhancements. It is very likely parents with the most money will almost always have the most successful children. The most enhanced people will be loaded with DNA from various sources cutting them off from their own ancestral family trees because they will not know what family tree they are perpetuating. This could cause people to experience significant loneliness and a sense of disconnection. Once we start genetic altering, the evil among us are sure to get into the act by taking revenge on unsuspecting others and/or producing ruthless leaders and the 'antichrist.'

McKibben argues against gene splicing from known or unknown donors but he is in favor of eliminating inheritable diseases in vitro via 'somatic gene therapy.'

The fear of nano-technology is about control or more importantly the lack or loss of control. What if self-replicating 'nanobots' escape and cannot be turned off? Nanobots that assemble water from oxygen and hydrogen in the air (useful in deserts), if out of control would choke us all to death by taking all the oxygen out of the air. Nanobots invented to destroy bioforms dangerous to humanity may attack and destroy all bioforms if out of control.

Cybernetics and artificial intelligence are developing rapidly and experiments are now implanting microchips in animal bodies and brains. Human insertions are just around the corner with some scientists already experimenting with their own bodies. There are critics of Mother Nature who unabashedly support all scientific progress and brave-new-world thinking. McKibben cites a speech by one such gentleman he aptly names 'Max More' at the Extropian Convention ('extropy' is coined as the opposite to 'entropy'). The speech was very critical of Mother Nature supporting all efforts to figure out things for ourselves and say goodbye to nature which has made many mistakes . . . the biggest of which is death. Why do we have to get old and die? Why do we have to be born out of a female body and why do we have to have a body at all with all its cells and blood? Why do we have to eat and defecate? Maybe science can fix our digestive tracts so we can take one tablet a month to survive. Maybe we can be born again out of something artificial instead of a natural body. Maybe science can download our brains into machines so we can linger around in cyberspace and never have to die. Dying versus immortality is the biggest question for science. Although we are not currently near becoming immortal our current scientific exploration is sure to increase our life spans soon. Research to find regenerative remedies to maintain fitness and stem declining organ function hit a milestone in 2011 when the aging process in mice via telomere integrity (A telomere is a repetitive region in DNA at the end of a chromosome which protects the chromosome from deterioration) was not only restored but reversed . . . in effect making the mice younger in health.[9] All of our myths and stories about immortality envision us either losing our souls and feeding on the blood of others (vampires) or becoming decaying horrors without eternal youth. G-d is immortal and humans die. If we try to change it and begin to worship technology and machines instead of Mother Nature and G-d we could end up in big trouble and/or evolve to a higher life-form.

If we were immortal, life becomes endless and we would become a different species. Stories and narratives would end . . . why tell them? Beginnings and middles would end because there will be no more endings. Literature and art is infested with mortality and our new immortal bloodless selves will not need it. Would we not get tired of endlessness, monotony, and the lack of a meaningful event? Would we sit around all day and contemplate problems such as where the universe came from and how it all started? To what end . . . for what purpose? Who exactly would the immortals be? The genetically wealthy? It certainly will not be the six or seven billion current earthlings or the ten billion projected for 2050. The divide between the 'haves' and 'have nots' in a world of immortal humans would dwarf any divisions we know today.

McKibben is an optimist who believes we still have choice and nothing is inevitable. Just because something is invented does not mean we have to use it. We invented the atomic bomb and after using it never used it again. The Japanese samurais rejected the use of guns and the Chinese abandoned advanced sea power. He believes that like the Amish, who examine each new technology and accept or reject it according to social and spiritual criteria, we can too. We must decide from the fullness of our present humanity, flawed as it may be.[10]

The beginning of the road to immortality may be with human organ transplants. Tens of thousands of people in America alone are registered and waiting for a genetically-suitable-somebody to die so that they can obtain their heart, kidney, liver, pancreas, or other organ. Although the problems with infection and rejection still loom large . . . the desire to live outweighs all the risks. Most Americans support organ donation but very few actually consent to it. American law allows families to override the wishes of the deceased decreasing the donor pool even more. Research is in full swing to create replacement organs from scratch using a patient's own tissues; i.e., harvesting one's organs. Biotech firms are pursuing making organs to order and have quietly fitted people with at least new bladders and maybe more. Healthy progenitor cells are extracted from the patient and placed in collagen scaffolding structure resembling the organ. Incubated with other nutrients the organ is harvested and placed in the patient's body where the collagen scaffolding is gradually reabsorbed into the body. There is no guarantee that what works with the harvested more complex animal organs (pancreas, heart, and lungs) will work with humans but it seems a better alternative to the transgenic practice of implanting animal organs into humans.[11]

Artificial Sensing

When it comes to the human senses technology has made great strides in sights and sounds in the last hundred years: high definition television, incredible still and video cameras, and surround sound. Technological smell advancements pale in comparison.

Chemical sensors can detect odor but artificial or electronic noses are far more efficient and versatile. The versatility of our sense of smell is challenging to replicate since it detects every type of odor in our environment; not just one as in detecting explosives at airports. The tens of millions of cells that make up the human olfactory system distributes a smell to 300-400 different types of receptors enabling us to detect tens of thousands of smells without having to have tens of thousands of different types of receptors. It is no easy task to replicate most things in nature and it is especially difficult to replicate our olfactory sense. Science is experimenting with an array of transistors made from various organic semiconductor materials which are very sensitive to slight changes in charge and bond structures. Although far from the olfactory capabilities of our nose the hope is that if the price can be decreased sufficiently devices can be built into food containers and pharmaceutical packaging, to indicate when a product is past its best. Progress is being made toward the possible installation of sensors in hospital and hotel ventilation systems to prevent the spread of infections (e.g., Legionnaires Disease).

Breathalysers have been developed to detect not only alcohol blood levels but lung cancer and other serious diseases. Systems have also been developed to emit smells and are being used in the treatment of war veterans suffering from post-traumaatic stress disorder (PTSD) by gradually reintroducing the patient to the environment that caused the stress and trauma in the first place thus increasing the realism of the experience until it no longer causes suffering. This is effective because our olfactory system is linked to the limbic system which controls memory; in the case of war veterans the smells of cordite, diesal, body odor, gunpower, and garbage bring back the memory of the war zone. Using smell memory to relax people and bring back fond memories has been used for decades: real estate agents relax potential buyers by making coffee before visiting a prospective home; shopping centers have cookie shops pumping out warm chocolaty smells making shoppers anxious to enter and smell more. Devices can now produce just about any kind of smell and mathematical algorithms are fully capable of mixing the odors. It is possible to sample a smell in one place and synthesize it with another. We have the technology to develop 'smellyvision' and a smelly Internet.

Devices to prevent the sense of smell from working are most helpful in preventing trauma from horrific memories like first responders on 9-11-01, or clean-up operations in the Asian tsunami of 2004 and Hurricane Katrina in 2005. The stench of decaying flesh combined with the odor of overflowing sewers is quite unnerving to say the least. It not only makes the work of relief workers difficult it can leave them with haunting memories for the rest of their lives. It is sometimes called the 'barbeque effect' because of the smell of burning meat can trigger horrific memories of burned bodies years later. There are products on the market like 'odorscreen' which is a gel that actually interferes with the sense of smell by releasing harmless compounds preventing our receptor proteins to perceive the intensity of malodors. As the iris controls the amount of light entering the eye our natural olfactory mechanisms prevent us from being overwhelmed with smell.[12]

For the sense of smell machines may match humans soon but to match our canine pets they have a much longer way to go. A dog's sense of smell is 100,000 times as sensitive as ours. Dogs are being used to sniff out gluten in food to protect people with celiac disease. Human tests of stool samples can detect cancer accurately only ten percent of the time while Labrador Retrievers can detect colorectal and bowel cancer with ninety-eight percent accuracy. There is a major push to utilize more dogs in Afghanistan to detect hidden roadside bombs. Exterminators are utilizing dogs to sniff out bedbugs with a ninety-five percent success rate. Rottweilers are helping scientists track endangered right whales at sea.[13]

The Internet

Since its inception in 1969 the Internet and other systems have matured. It has been tweaked in piecemeal fashion by engineers over the last four decades to accommodate billions of users. It is not without challenges: spam, viruses, denial of service attacks, accommodating an increasing array of new devices (cars, mobile phones, wireless sensors, etc.). Initiatives are underway, based on our accumulated knowledge of what works and what needs to be improved, to develop an entirely new Internet system to accommodate projected future needs. The visions of the future assume the number of devices will continue to multiply in unforeseen ways with chips and sensors embedded in almost everything . . . ergo hundreds of billions of devices worldwide. The vast majority of future traffic is assumed to be wireless/mobile machine to machine, not wireless/mobile human to human. One idea is called 'trust-modulated transparency' in which the network's traffic-routing infrastructure becomes like a security guard in an office building judging the trustworthiness of packets of data that pass by. Dubious packets are re-routed for screening and possibly blocked. Traffic flows freely between devices that trust each other and are closely scrutinized between devices that do not trust each other. Another idea called 'internet indirection infrastructure' would overlay an additional addressing system on top of the internet-protocol numbers now used to identify devices on the Internet thus making it easier to support mobile devices and allowing 'multicasting' of data to many devices at once: this idea would enable the efficient distribution of audio, video, and software. These ideas of re-designing the Internet are being tested in laboratories that allow many such prototype networks to operate at once without interfering with each other. Diagnostic systems are also being explored that can determine what caused a breakdown such as a failed videoconference: was it the Internet provider or poorly configured equipment? Today's dumb network does not do diagnostics . . . a newly designed network would have to collect and share performance measures.

Today's dumb network was a deliberate design decision making it as simple as possible thus making it a vibrant platform for innovation. New applications emerged: the Web, peer to peer file sharing, and Internet telephony. Although these applications were not precluded by the original design they were not explicitly provided for. The Internet blindly passes packets of data between devices because it is dumb . . . on the edges of the network these packets can be upgraded to do innovative things. The non-discriminatory Internet (known as the 'end-to-end-principle') is one of the most highly regarded aspects of the Internet's design. Innovations at the edges is promoted but core innovations are prevented. 'Active networks' or 'metanets' has potential for core innovation. Routers (boxes that direct Internet traffic) could be replaced with more flexible devices that are able to learn new communications protocols when needed and devices on the edges could dynamically reprogram all the routers to use whatever new protocol they wanted. Routers could be enabled to partition themselves internally thereby not affecting other users of the network. The advantage is the creation of new protocols for video streaming, file sharing, and new things not yet created. The disadvantage is that routers are fast because they are single minded and adding complexity to them might make them more versatile with less than adequate performance.

Metanets provide an answer to the 'clean-slate' approach to changing the core of the Internet. It would allow multiple internets to run in parallel with competition between different protocols; e.g., if one protocol was more secure, security-sensitive users would have reason to use it. Engineers working on these new ideas and devising new protocols takes time confronting technical hurdles in addition to social, political, and economic arguments against implementation. The current Internet architecture allows innovation, promotes free speech, allows spam, and allows illegal downloads. To change the core of the Internet (i.e., its 'soul') requires making choices with very wide implications. Many believe the greatest impediment to architectural design is not technical but the political disagreements over how it should work.[14]

Energy use

'Standby power' by our electronic devices consumes a lot of energy. A typical microwave oven consumes more electricity powering its digital clock than it does heating food even though it takes 100 times more power to heat food than to run the clock but it stands idle 99% of the time. DVD players, stereos, computers also stand idle most of the time consuming a huge amount of energy collectively. We have moved from an electromechanical world that was on and off to an electronic world that is never off! The state of California was the first public entity to require mandatory standby requirements for various electronic devices in 2006 . . . the first such obligatory regulations in the world. In a 1998 study it was estimated that standby power accounted for approximately 5% of the total residential electricity consumption . . . wasted energy equivalent to the output of eighteen power stations. Additional studies conducted in America, France, Netherlands, Australia, and Japan estimate the standby consumption of electricity in ranges from 7 to 13%. Unfortunately there is very little incentive for manufacturers to make devices with low standby-power consumption. The worst offending devices consume twenty watts in standby mode and it is estimated that all the standby functions can be performed with one watt or less. Reducing power demands, reducing electricity costs to the consumer, and reducing carbon-dioxide emissions are very clear positive outcomes for reducing standby power consumption.

The older power supplies contain iron cores surrounded by windings of copper wire. Today 'switch mode' technology is available to convert a main's electricity to low voltages used in small devices. With the addition of extra circuitry it can reduce standby mode consumption to a fraction of one watt. The old inefficient power supply devices are cheaper and since the cost of more energy consumption is borne by the consumer the manufacturers have no incentive to change to 'switch mode' technology. Regulators around the world have encouraged manufacturers to voluntarily make their products less power hungry. Australia and South Korea have been the only countries to adopt a voluntary national 'one-watt' standard aiming to fully meet this standard for all new products sometime in the future. After the 2001 California energy crisis President George Bush issued Executive Order 13221 requiring that every government agency "when it purchases commercially available, off-the-shelf products that use external standby power devices, or that contain an internal standby power function, shall purchase products that use no more than one watt in their standby power consuming mode." The counter argument by the American electronic industry is two fold: 1. that imposing mandatory regulation will stifle innovation and limit consumer choice; 2. industry-led standards are a much better approach for focusing on energy-related efficiency than standards set by governent.[15]

Security, Biotechnology, and Neuro-technology

The number of attempted and carried out terroristic attacks seem to be increasing world wide. Technology is playing large from the fear of suitcase atomic bombs, to using printer cartridges loaded with explosives to blow up cargo airplanes. A new kind of warfare called micro-terrorism or small-scale terrorism has emerged in which practitioners choose operations most likely to succeed instead of large spectacular ones. These small-scale terrorists are driven from the local level and at their core are not conducted from on high but rather bubbles up from below in an effort to bleed the enemy to death. Operation Hemorrhage which was behind the printer-cartridge attack cost a total of $4,200 involving two phones, two printers, and shipping costs. In the 1990's Al-Qaeda managed dozens of operatives in several countries and planned for simultaneous bombings like the embassies in Kenya and Tanzania and significant targets like the U.S.S. Cole. All this culminated in the 2001 attacks . . . simultaneous and very significant targets. Now the focus is on disrupting Western life through a series of much smaller attacks. Micro-terrorism is counting on the democratization of technology that is sweeping the world. Power is shifting from large institutions to motivated individuals. Technology allows people to leverage the weight of institutions against themselves. Being small insures that it is hard to detect and control. The enemy is hard to find in Afghanistan, Yemen, and Somalia and the strategy quickly changes to stabilizing the country . . . i.e., nation building. These countries have never been orderly or stable and all the enemy has to do is hide and mail bombs or send individuals to blow themselves and others up. If the terrorists are Western citizens with no previous track record of death and destruction the prevention gets extremely more complicated. The democratization of technology, access, information, and the Internet are all leading to the democratization of violence.[16]

Many of the technologies that are racing ahead with incredible speed are replacing human workers in factories and offices with machines. They are making stockholders wealthier and workers poorer. In fact they are accentuating the existing inequalities in the distribution of wealth. The technologies of lethal force (i.e., war) are very profitable. The ethical unanswered question for science is whether the job the technology is designed to do is actually worth doing? Inherently safe nuclear reactors to generate electricity were worth doing before the technology became entangled with politics and bureaucracy and it became clear that nuclear energy was not and never could be the great equalizer. The technology of CAD-CAM, computer-aided design and computer-aided manufacturing has greatly benefited many manufacturing and architectural firms; it has been definitely doing the job it was designed to do and has been very worthwhile doing. Will the future replace CAD-CAM with CAS-CAR, computer-aided selection and computer-aided reproduction? A CAS-CAR world would enable you to program the pet you want by color and behavior, transmit the program to the artificial insemination laboratory for implementation, and then pick up your pet with satisfaction guaranteed by the software company. Designing cats and dogs this way is not like designing houses or machines ... it is an ethically dubious business. If the future allows the design of our pets how far off is the CAS-CAR software to design our own babies. The first generation of CAS-CAR could be in the form of three-dimensional printing. Consumers can submit digital designs of products to online services that print the object and ship it back to the customer. Three-dimensional printers use successive layers of different polymers that gradually spray until a 3-D object is created. The technology has gone from polymer prototypes to actual working gears and machines such as airplane landing gears and grandfather clocks. Medical implants, jewelry, lampshades, racing-car parts, solid state batteries, and customized mobile phones are also being produced. In 2011 more than twenty percent of 3D printers outputs were final products rather than prototypes.[17]

The ways science can work for good or evil in human society are many and various. Generally with exceptions science works for evil when its effect is to provide toys for the rich, and works for the good when its effect is to provide necessities for the poor. Cheapness to produce is an essential virtue. Nuclear energy worked mostly for evil because it remained a toy for rich governments and rich companies to play with. Toys are not about playthings but about technological conveniences that are available to a minority of people and make it harder for those excluded to take part in the economic and cultural life of the community. Necessities for the poor include food, shelter, adequate public health services, adequate public transportation, and access to decent education and jobs. Many scientific advances in the last hundred years have been social equalizers narrowing the gap between rich and poor rather than widening it (electric light, telephone, refrigeration, radio, television, synthetic fabrics, antiboiotics, vitamins, and vaccines). The strongest efforts in applied science more recently have been concentrated on products that can produce profit. Since the rich can pay more than the poor, in our market-driven economy it can result in toys for the 'haves' not the 'have nots' (laptop computers, iPhones, iPads). Pure scientists have become detached from the mundane needs of humanity and applied scientists have become more attached to immediate profitability.

The 21st century contains three new ages: Information, Biotechnology, and Neuro-technology. The Information Age is here big time via the Internet and the incredible availability of information worldwide. The Biotechnology Age is developing rapidly with DNA sequencing and genetic engineering. The Neuro-technology Age is in its infancy developing neural sensors, and exposing the inner workings of human emotion, and personality to manipulation. These three new Ages offer wealth and power to the people who possess the skills to understand and control them. They are likely to bypass the poor and reward the rich. The widening gap between technology and human needs can only be filled by ethics which can be a force more powerful than politics and economics. The environmental movement has concentrated its attention upon technological evils (nuclear power, global warming, global deforestation, etc.) rather than the good technology has failed to do. Religion like science has had and still has prophets of doom and prophets of hope. The prophets of hope, as in the religious context, need to predominate in science towards a better world and not the end of humanity as we know it. As we learn more about biological processes that affect human beings which will lead us to new technologies we need to use this knowledge to prevent tragedies and ameliorate the human condition. Our species may not be perfectible but we are certainly capable of improvement.

In biotechnology the idea of artificially improving the human race is widely debated but the science is progressing full speed ahead. Human improvement, regardless of the effort to stifle it by regulation or law, will be widely practiced. The freedom of people to choose their own improvement cannot be permanently denied (better health, longer life, improved disposition, greater physical strength, more intelligence, etc.). Our loyalties, fears, hatreds, economic injustices, and social injustices all grew with the human species over many generations and have deep historical roots. Any change whether it be in social norms, political norms, or in science brings about conflict between the old ways and the new. Our new technologies are dangerous and liberating. We can only predict the future by examining the past and the past clearly demonstrates that in the long run social constraints will bend to the new realities. Self improvement will continue to motivate all of us . . . the question is to what end will it be utilized?[18]

REFERENCES TECHNOLOGY

1. Gelemter, David Hillel. (1992). *Mirror Worlds, or, The Day Software Puts the Universe in a Shoebox: How it will Happen and What it will Mean.* New York: Oxford University Press.

2. *The Economist* (November 4, 2010). "The real and the digital worlds are converging, bringing much greater efficiency and lots of new opportunities, says Ludwig Siegele. But is it what people want?"

3. *The Economist* (November 4, 2010). "A special report on smart systems . . . A sea of sensors . . . Everything will become a sensor – and humans may be the best of all."

4. O'Reilly, Tim. (June 25, 2010). "Web Squared: Why the sensor web is the next big thing in Web 2.0" *http://bit. ly/V3B8Z.*

5. *The Economist* (February 25, 2010). "Show Me . . . New Ways of Visualizing Data."

6. Halpern, Sue. (June 10, 2010). "The iPad Revolution" *The New York Review of Books.*

7. Lewontin, Richard C. (June 21, 2001). "Genes in the Food!" *The New York Review of Books.*

8. Rissler, Jane, Mellon, Margaret. (2001). *The Ecological Risks of Engineered Crops.* Boston: MIT Press.

9. Jaskelioff, Mariela; Muller, Florian L.; Paik, Ji-Hye; Thomas, Emily; Jiang, Shan; Adams, Andrew C.; Sahin, Ergun; Kost-Alimova, Maria; Protopopov, Alexei; Cadiñanos, Juan; Horner, James W.; Maratos-Flier, Eleftheria; DePinho, Ronald A. (January 6, 2011) "Telomerase reactivation reverses tissue degeneration in aged telomerase-deficient mice" *NATURE.*

10. McKibben, Bill. (2003). *Enough: Staying Human in an Engineered Age.* New York: Times Books.

11. *The Economist* (March 9, 2006). "Organs to order Biotechnology: Could the creation of replacement organs, grown to order for particular patients, be just around the corner?"

12. *The Economist* (March 9, 2006). "What the nose knows . . . Smell technology: Technology can manipulate and reproduce sight and sound with amazing fidelity. But what about smell?"

13. Ellison, Jesse (February 13, 2011). "Are Dogs Stealing Our Jobs?" *Newsweek.*

14. *The Economist* (March 9, 2006). "Reinventing the Internet: Networking: New initiatives aim to overhaul

the internet. But how can a "clean slate" redesign ever be implemented?"

15. *The Economist* (March 9, 2006). "Pulling the plug on standby power . . . Energy: Billions of devices sitting idle in "standby" mode waste vast amounts of energy. What can be done about it?"

16. Zakaria, Fareed. (December 15, 2010). "Essays . . . The Year of Microterrorism" *Time Magazine*.

17. *The Economist* (February 12, 2011). "Three-dimensional printing from digital designs will transform manufacturing and allow more people to start making things."

18. Dyson, Freeman (April 10, 1997). "Can Science Be Ethical?" *The New York Review of Books*.

HOLLYWOOD AND SCIENCE FICTION

Introduction

Science fiction movies like science fiction literature uses our imagination to expand our reality to wherever we want to go. It is about entertainment but may have more realistic implications if we understand that our imaginations come from our thoughts which are also part of our universe; i.e., the non-physical part. It is through our thoughts that we can 'go where no man has ever gone before' and travel faster than the speed of light; i.e., arrive at a very distant location light years away instantaneously in our minds. It does not mean that all thoughts about life and the future are correct nor does it mean we can accurately predict the future. What it means is that our thoughts, as ridiculous and incredible as they may be, can give us insight into present and future possibilities. I have selected a number of popular Hollywood science fiction movies that give us insight about our species, our technology, and the future of a technologically advanced Homo Sapiens.

2001: A Space Odyssey

The first blockbuster Hollywood sci-fi film goes back to 1968: "2001: A Space Odyssey" directed by Stanley Kubrick. It is considered a science fiction classic about the exploration of the unknown. It prophetically showed the enduring influence computers would have in our daily lives. The movie portrays how man is dwarfed by technology and the vastness of space. Much of the film is in dead silence (accurately depicting the absence of sound in space) or the sound of human breathing in a spacesuit. The film is about space travel, discovery of extra-terrestrial intelligence, and advanced technology. The film opens with the camera panning upward from the cratered surface of the moon in the foreground. The perspective is from behind the moon and in the distance is the sun rising over the Earth-crescent in the vastness of space. The scene shows the Earth, moon, and sun in a vertically symmetrical alignment or conjunction; it is later revealed that a monolith was buried on the moon, possibly at the moment of this conjunction. The film is divided into four episodes:

1. "The Dawn of Man" in which a primeval ape man makes a breakthrough by becoming endowed with intelligence after experiencing a mysterious black

monolith. In the year 2000 a lunar journey discovers a similar monolith on the lunar surface sending its signals to Jupiter.

2. "Jupiter Mission" during which an eighteen month journey to the planet Jupiter takes place in the year 2001.

3. "Jupiter and Beyond the Infinite" involves a mystical experience in another time and dimension.

"The Dawn of Man" opens in the Pleistocene era . . . four million years ago at the location where the human race itself, evolving from primitive apes was born. A tribe of ape-men appear (Australopithecines) eating grass; they have not yet developed the means or tools necessary to become carnivores like other predators. They scrape out an existence in their ongoing struggle to survive against competitors (other tribes), and carnivores that eat them for breakfast, lunch, and dinner. They huddle together at night as they live and sleep in fear. At the dawn of the second day a tall black monolithic slab (the first monolith) appears in their cave with an eerie humming sound (symbolic of the religious and spiritual unknown). The ape-men react nervously but eventually approach it cautiously. The leader of the clan extends his finger to touch the monolith approximating the Biblical equivalent of eating the forbidden fruit of Knowledge. All the ape-men huddle around the monolith as another celestial alignment occurs. Later that day the leader is foraging for food and comes upon the bone of a ravaged antelope. As the celestial alignment with the monolith occurs the ape-man realizes that the bone can function as a weapon as he smashes other bones with it. Very shortly thereafter he becomes a carnivore killing and sharing a fallen tapir with his tribe. The monolith has mysteriously gifted the man-ape with a transition to a higher order of intelligence with an ability to reason and the power to use tools for killing. The man-ape has extended his reach . . . the beginning steps toward humanity. The next day the carnivorous man-ape tribe is able to drive the weaponless (and tool-less) neighbors away as they swing their bone-tools and kill one with a deadly blow to the head. This is humanity's first bloody war presumably alluding to the story of Adam's son Cain killing his brother Abel. The 'enlightened' man-apes gain dominion in the animal world and take an evolutionary step toward humanity. Realizing this incredible step the leader of the pack flings his piece of bone high in the air in celebration. It flies upward turning end over end.

The lunar journey occurs in the year 2000 or four million years later with no sub-titles separating the lunar journey from the "Dawn of Man." It seems to imply that man in both eras (the Australopithecine and the Space-Age Man) is essentially the same aggressive creature with savage impulses who has successfully survived in another hostile environment (space). The tossed bone weapon instantly rotates and dissolves into an orbiting space satellite from Earth . . . a technological tool/weapon (orbiting nuclear platform) or machine from another era that was ultimately derived from the first tool-weapon. The bone toss is a metaphor for the launch from Earth toward the moon and all the technological advances that have occurred. A Pan American commercial space shuttle Orion soars from Earth to the moon via Space

Station 5 (ironically Pan Am went bankrupt in 1991). Pilots guide the tubular shuttle (a symbol of a phallus) into the spoke hub of the circular revolving space station (a symbol of an egg) while a stewardess tends to the passengers. The entire docking procedure is symbolic of a copulation. In the space station corporate logos are visible in the long entryway (Hilton, Howard Johnson's Restaurant, and Bell Telephone). The journey from the space station to the moon base is once again imagery of reproductive life . . . a hatch opens under the landing zone and gently brings the spacecraft into its interior. The round impregnated 'ova' implants itself into the 'uterus' of the mother. The American moon base Clavius is under conditions of highest security because a second monolith has been discovered on the moon (another indication of extra-terrestrial life and their desire to provide further guidance to mankind as it also exerts its unmistakable will on human beings in a different era). An alternative 'cover story' has been fabricated about a possible epidemic at the base. The government fears that any leak of the discovery may cause anxious panic and disorientation among the families of Clavius personnel. It has become quite obvious that the monolith was deliberately buried four million years earlier with the same eerie humming sound of Earth's monolith. There is a scene of select scientists at the monolith awed and stirred by this first view of an alien form. The leader touches it with his thick glove as the fourteen day lunar night ends and the sun, moon, and Earth have formed a conjunctive orbital configuration. The object emits an ear-piercing electronic screeching noise once it is touched by the sunlight; the solar powered machine functions as a radio signaling device aimed at the planet Jupiter. It has alerted the aliens that buried it that man has reached another more advanced level of consciousness and intelligence.

The "Jupiter Mission" occurs eighteen months later in 2001, the first eighteen-month 500 million mile manned mission. Their important mission is to follow the path of the radio signal sent to Jupiter and to find the origin of the alien culture that planted the monolith on the moon and and/or caused the unexplainable radio transmission. The enormous spaceship the Discovery has a shape similar to the man-ape's bone. The spaceship also resembles a half-developed fetus floating in the amniotic fluid of space. Some of the astronauts are hibernating in pods ready to be born . . . or awakened. Despite the incredible importance of the mission, life onboard is very boring and monotonous. The two of the five astronauts who are not hibernating do not communicate with each other in a parody of many earthly relationships. The spaceship is controlled and monitored by the sixth member of the Discovery crew HAL 9000, an even-toned, talkative, alert, 'thinking' and 'feeling' super computer. HAL maintains the electronic systems and the humans, bored with their routines are at his mercy. HAL is a perfect technological power that can reproduce most of the human brain's capabilities (superhuman traits, murderous thoughts, and insanity). HAL who could reproduce (experts preferred the word 'mimic') most activities of the human brain with greater speed and reliability was not concerned about the problems with hibernating astronauts. The name HAL is derived from an acronym from the words 'H'euristic 'AL'gorithmic . . . two basic types of learning systems. Only

HAL knows the real mission of the trip and like the cover story on the moon the five astronauts are unaware of the real purpose. HAL who is interviewed with the two other astronauts by a British Broadcasting Company (BBC) reporter provides more detailed, clearer, more concise answers that appear to be more human. When asked about his very significant responsibilities on the mission and if he ever has any doubts or lack of confidence HAL replies: "Let me put it this way, Mr. Amer The 9000 series is the most reliable computer ever made. No 9000 computer has ever made a mistake or distorted information. We are all, by any practical definition of the words, foolproof and incapable of error."

The emotionless astronauts appear bored with the drudgery of their technological routine and probably from being in the company of an omniscient, controlling, expressionless computer. The men's deep trust in HAL's infallibility has destroyed their will and vitality. HAL's concern about 'odd things' in the mission may be a sign that the 'perfect machine' is failing or showing signs of diminished responsibility . . . the first thing to break down. The error-detection systems fouling up may be from the computer's role in keeping vital secrets about the mission from the crew . . . especially when they start asking HAL questions about the 'true' mission. The fault detected is in the communications system AE35 . . . a space walk repairs the problem by the insertion of another AE35. Diagnostic tests are run on the supposedly defective unit and it works just fine. HAL is puzzled and calmly recommends: "I would recommend that we put the unit back in operation and let it fail. It should then be a simple matter to track down the cause. We can certainly afford to be out of communication for the short time it will take to replace it." The twin HAL 9000 on earth indicates that the spaceship made an error predicting the breakdown of the AE35 unit . . . something Mission Control thinks is impossible to happen but it has. Is the 'infallible' tool created by man deliberately conspiring against its creators? Is HAL's crack-up the result of inborn (programmed) human error? Since HAL is a thinking machine assuming human characteristics is HAL becoming paranoid, threatened, and fearful that the end of the Jupiter mission would mean its own demise, disconnection, and extinction?

The two astronauts truly concerned about HAL, attempt to talk about it in a sound-proofed, sealed pod so HAL cannot hear them but can see them. They plan to replace the original AE35 unit: if it does not fail as HAL predicted HAL would clearly be at fault and must be disconnected. Through the viewport HAL has been able to read the lips of the astronauts' conversation and formulates his own counter-plan in reaction to their agenda. The original AE35 unit is not reinstalled because HAL uses the mechanical claw-arms to snap the astronaut's oxygen lines (a symbolically snapped/severed umbilical cord) killing him. Dave, the surviving non-hibernating astronaut knows HAL has killed his companion when he asks what happened and HAL replies: "I'm sorry Dave. I don't have enough information." To cover up any evidence of his own error HAL vengefully proceeds to destroy the hibernating occupants of the spaceship by disconnecting them and then refuses to allow Dave back into the spaceship. Dave, like the man-ape uses his unique human tool of intelligence to outwit HAL in a life and death game of strategy that allows him to evolve to the next level. He uses the explosive bolts on his pod's hatch to eject himself out and into Discovery's emergency door. He then releases the oxygen into the chamber and survives (another startling image of reproductive birth). Dave immediately goes to the 'brain room' as HAL promises to do better and knows he has made mistakes but has fully recovered and is back to normal to proceed with the mission. As logic does not stop Dave, HAL turns to begging and pleading while admitting he is afraid. Dave de-brains, lobotomizes, dismantles and disconnects HAL's higher-logic functions. HAL's pleading continues as his 'mind' gradually decays and goes through a second childhood and infancy before reaching senility and becoming imbecilic. The computer maintains its calm tone but still expresses a full range of genuine emotions while dying. As the ship enters Jupiter's space it triggers the playing of a pre-recorded televised briefing that tells the truth about the mission and the meaningless journey now has relevance with the one surviving astronaut.

In the final chapter of the film "Jupiter and Beyond the Infinite" man must find an alternative now that technology has failed him. Dave completes the flight and finds the life-source of the Universe. He is completely human and vulnerable without the crippling dependence on a machine. The sun, Jupiter, its many moons, and Discovery line up with another monolith (the third) that hurtles through space toward Jupiter's moons. Dave pursues the monolith in his pod and is sent down a tunnel of light speckles (a time warp termed the Star Gate) moving beyond the speed of light. During his transcendental journey and space odyssey he shakes and watches in wonder at the cosmic whirlpool rerouting him toward other dimensions with unimagined speed. During his passage (through the birth canal) he is transfigured (reborn) into a higher form of intelligence or universe of evolutionary life on his way to the alien planet. On the journey through alien solar systems extreme close-ups of Dave's facial features and dilated eye reveal the patterns of the Universe that he has become part of. There are explosions of nebula, swirling gases, bursting constellations, bright stars, blazing skies, images of swimming sperm, colorful and desolate landscapes. He lands in a new realm of physical reality and he lands in semi-familiar surroundings from his own subconscious memories. In the end he comes to a cosmic ornate suite/bed chamber. Dave ages . . . his second stage of rapid regressive (and progressive) transformation. He enters the marble bathroom to find in the mirror that his life span is rapidly passing by. He looks back to the bedroom to find another reincarnation of himself. He sees himself a third time in the dining room as an even older gentleman is clicking his eating utensils on the table as he eats an elegantly served meal. He continues to dine with wine and bread (a last supper with sacramental elements). In the fourth stage of rapid aging he sees himself as a bald dying man, lying on the bed, looking to be one-hundred years old and shrunken in size. He slowly and feebly reaches his trembling hand out toward another glowing and mysterious monolith (the fourth) that appears at the foot of the bed. As he reaches he presumably dies and is transformed (evolved and reborn). He dissolves into a glowing, hazy, translucent fetus or embryo in utero that rests on the bed.

The black monolithic slabs symbolize the alien beings who decide that he should be reborn. The final sequence visually portray the many reproductive allusions seen throughout the movie: procreation, gestation, birthing, and nursing. He is assisted by the aliens to make a symbiotic change in consciousness toward a more completely civilized human being with a universal knowledge of existence. The end result of the space odyssey is not a greater more infallible machine, but a greater, more fully-realized being produced in a second childhood. A zooming close-up of the towering black monolith at the foot of the bed takes us back into the blackness of dark space. Dave re-emerges in the embryo with his own serene, wise features. He becomes reborn as a cosmic, innocent, orbiting 'Star-Child' that travels the Universe without any technological assistance. The last image of the film is of a large, bright-eyed luminous embryo in a translucent uterine amnion or bluish globe. An evolved reborn superhuman floating through space.

The film takes us through the putative evolution of the human race from ape to man to spaceman to angel/star-child/superman. Evolution has been outwardly directed toward another level of existence . . . from isolated cave dwellings to the entire Earth to the moon, to the solar system and to the Universe. "2001 A Space Odyssey" is hopeful and optimistic about humankind's potential for the future. We may create incredible powerful machines that will threaten our existence and evolution but in the end 'we shall overcome.'[1]

Jurassic Park

Jurassic Park is a 1990 science fiction novel by Michael Crichton made into a movie by Steven Spielberg in 1993. Unlike the forward evolution of humans and machines in "2001: A Space Odyssey" it is considered a cautionary tale on unconsidered biological tinkering in the same spirit as Mary Shelley's "Frankenstein." Using the mathematical concept of 'chaos theory' and its philosophical implications it genetically recreates dinosaurs for amusement and profit (i.e., Jurassic Park is an amusement park located 120 miles off the coast of Costa Rica that never opens). The park is owned by billionaire John Hammond whose investors are concerned about the park's viability. Hammond brings three noted scientists to the park to act as consultants (paleontologists Grant and Sattler and mathematician-chaos theorist Malcolm) along with lawyer Gennaro who represents the nervous investors. The paleontologists are fresh and a hopeful counter-balance to pessimists Gennaro and Malcolm. Malcolm, who was consulted during the park's creation is emphatic in his prediction that the park will collapse, as it is an unsustainable simple structure bluntly forced upon a complex system.

Jurassic Park showcases cloned dinosaurs which have been recreated using damaged DNA found in mosquitoes that sucked dinosaur blood and then were trapped and preserved in amber. Gaps in the genetic code have been filled with reptilian, avian, or amphibian DNA. All the island's clones are bred to be female and lysine-deficient. The four visitors are grouped with Hammond's two grandchildren (Tim and Lex) for a tour. They find a Velociraptor eggshell to prove that they are breeding as Malcolm had predicted: "Life will find a way!" The population graphs the genetics introduced were normally distributed reflecting a breeding population rather than displaying the distinct pattern that a population reared in batches ought to display. The park's motion detectors were set to search only for the expected number of creatures and not for any higher number thus not knowing if any procreation was taking place. Malcolm also points out the height distribution of the Procompsognathus forms a Gaussian distribution (bell curve or normally distributed), the curve of a breeding population. The software programmer that controls the park, Dennis Nedry steals fifteen frozen embryos (one for each of the park's fifteen species) during his orchestrated security system shutdown. Nedry has been promised untold riches by a competing corporation if he can deliver the embryos. During a sudden tropical storm while trying to deliver the precious cargo, he is killed by one of the dinosaurs. Dependent on Nedry for controlling the computerized park, his absence leaves the security system down (electrified fences become deactivated) and the dinosaurs begin to escape attacking vehicles, buildings, and killing and injuring people. The park's upper management struggle to return the park to working order which they do for a time but forget that they have been running on auxiliary power which runs out shutting down the park for a second time. Grant manually gets the power back while Tim is able to contact the supply ship Anne B to tell them to return. Gennaro tries to order the island destroyed but Grant rejects his authority because they have a responsibility to learn what happened and how many dinosaurs have escaped to the mainland. Hammond is contemplating how he can make another improved park without repeating his past mistakes before he is killed and eaten by a pack of compys. It is revealed that the frog DNA used to fill gaps in certain strands enabled some of the dinosaurs to change sex, as some species of frogs can do. In the conclusion the island is violently destroyed by the fictional Costa Rican Air Force. Survivors of the incident are indefinitely detained by the United States and Costa Rican governments. Grant is visited by an American doctor who lives in Costa Rica who tells him that an unknown pack of animals have been eating crops rich in lysine (the molecule in which the animals were designed to be deficient) and killing chickens as they migrate toward the Costa Rican jungle. The doctor also tells him that none of them with the exception of Tim and Lex will be leaving anytime soon.[2]

Gattaca

The film "Gattaca" was written and produced by Andrew Niccol in 1997. Vincent Freeman is one of the last 'natural' babies born into a sterile, genetically-enhanced world where life expectancy and disease likelihood are ascertained at birth. Myopic with a congenital heart defect he is predicted to die at age thirty. He can only assume menial jobs in a society that discriminates based on your DNA score and not your gender, race, or religion. His dream is to travel in space. Gattaca Coporation would never take a low scoring specimen like Vincent. He contacts a very high scoring genetic specimen who won a silver medal (not a gold medal) in a very high-profile competition and became a paraplegic when because of his failure to be first he attempted suicide by jumping in front of a car; again not achieving his goal. Since no one knows of the accident that occurred overseas Vincent is able to 'buy' through an unlawful underground financial transaction the DNA of Jerome Morrow whose name he assumes. He requires orthopedic surgery to increase his height, contact lenses to replace his glasses and match Jerome's eyes, and constant practice to favor his right hand over his left. He can pass any DNA test (blood, tissue, and urine samples) by using his 'valid' DNA taken from Jerome. He must take extra precautionary steps to insure he leaves no traces of his real DNA 'in-valid' identity by cleaning his desk of dead skin droppings and insuring no hair of his body is left anywhere at work. Born for a life of scorn and pity he is now the perpetrator of an incredible fraud that could lead to huge fines and jail time . . . he is now a heretic against the new order of genetic determinism . . . or in a new twist to an old word . . . a "de-gene-erate."

He learns how to deceive DNA and urine sample testing. Gattaca Corporation, the most prestigious space-flight conglomerate of his time selects him for a manned mission to Saturn's 14th moon Titan. Through his incredible determination he becomes the company's best celestial navigator. By passing all gene tests through the utilization of Jerome's hair, skin, blood, and urine his dream is within reach until his program director is murdered one week prior to his scheduled launch. He carelessly has lost an eyelash in the building which is turned over to the police and tests reveal that it belongs to a Vincent Freeman who they clearly want to question about the murder. The police begin extra searches and new gene tests at Gattaca to catch the murderer or at least bring in Vincent Freeman for questioning. Because of his fraudulent high DNA score women are very attracted to him and he pursues his coworker Irene. Not knowing who he can trust and about to lose everything (Irene, flying in space, his prestige, and his freedom) the police close in as his launch date nears. After numerous close calls the Director of Gattaca, who wanted to buy time for the launch because the window of opportunity comes once every seventy years, is arrested for the murder of the mission director.

One of the detectives who is Vincent's estranged brother, Anton confronts him. Anton tries to convince Vincent to come with him for protection before he is found out. Anton is acting more out of his own insecurity and sibling rivalry because he can not understand how Vincent was able to get the better of him despite his genetic superiority. They settle their competition in the same fashion as when they were children, by seeing who could swim out the farthest in the ocean. As in childhood Vincent wins and once again saves his brother from drowning. The environment and expectation of their 'brave new world' made Anton so competitive that he refused to save any strength to swim back since he was willing to risk everything to win and succeed. Vincent conversely concerned himself with swimming out and back again and therefore saved his strength; his realistic fears kept him from harm and truly testing his limits.

As the launch day arrives Jerome says goodbye to Vincent and tells him he intends to travel also. He leaves Vincent with enough genetic samples to last for many years. Vincent is very grateful and thanks Jerome for lending him his identity that has allowed his success and dream fulfillment. Jerome tells him that it is he that should be grateful since Vincent lent Jerome his dreams. On the way to the launch site Vincent is stopped for one more DNA test; since he is without Jerome's samples the test reveals his 'invalid' status. The doctor knows of Vincent's true identity since he has seen him favor his left hand . . . something a right-handed man would not do. The doctor alters the test result allowing Vincent to proceed. The doctor reveals that his son admires Vincent and wants to be an astronaut just like him in spite of an unforeseen genetic defect that would already rule him out. As the launch takes place Jerome is shown incinerating himself wearing his sliver medal which turns gold in the flames.[3]

"Gattaca" is about the power and force of will, spirit, desire, and intention. High scoring Jerome fails to succeed despite having every natural advantage. Low scoring Vincent transcends his deficiencies through pure force of will, spirit, desire, and intention. Being born of privilege clearly does not always translate into worldly success. Many children of wealthy families and very successful parents struggle throughout their lives. Sometimes a high expectation of success can be the very deterrent to achieving it. Jerome's expectation may have been so high based on his incredibly high genetic score that it may have made him less hungry to achieve his 'golden' goal. Jerome may have had a loss of desire since his expectation of himself was so high that he may have consciously or unconsciously thought he had it made and did not need to worry. The world portrayed in the film adversely affects the humanity of its characters. Human life is not based on any score . . . it is based on being true to one's self and to others and doing the best one can with whatever G-d has provided us. Albert Einstein and Abraham Lincoln who both had some significant genetic abnormalities would have been excluded in the world of "Gattaca."

The Terminator

"The Terminator" is a 1984 science fiction action film co-written by James Cameron, Gale Anne Hurt, and William Wisher Jr. The Terminator is a cyborg assassin sent back in time from the year 2029 to 1984 (Los Angeles) to kill Sarah Connor. Kyle Reese is a soldier from the same future sent back in time to protect Sarah.

In post-apocalyptic 2029 artificially intelligent machines seek to eliminate what is left of humanity. The Terminator kills two other Sarah Connor's he finds in the telephone directory before attempting to kill the real future mother-savior of humanity in a nightclub. Kyle is able to protect Sarah from harm and they both escape with the Terminator in hot pursuit. While escaping in Sarah's car Kyle tells the unbelieving Sarah about the future and time travel. He describes the Terminator "as an infiltration unit, part man, part machine. Underneath, it's a hyper-alloy combat chassis, microprocessor-controlled, fully armored. Very tough . . . but outside, it's living human tissue. Flesh, skin, hair . . . blood. Grown for the cyborgs." Sarah bites his arm hard in an unsuccessful effort to escape as Kyle coldly tells her: "Cyborgs don't feel pain. I do. Don't . . . do that again!" He goes on to explain that in the near future (20th century) an artificial intelligence network called Skynet will become self-aware and initiate a nuclear holocaust of mankind. Kyle explains:

> ". . . everything is gone. Just gone. There were survivors, here, there, nobody knew who started it. It was the machines. Computer Defense Network, new, powerful, hooked into everything, trusted to run it all. They say it got smart . . . a new order of intelligence. Then it saw all people as a threat, not just the ones on the other side. Decided our fate in a microsecond . . . extermination!"

Sarah's yet to be born son John will rally survivors and lead the resistance movement against the machines. As the Resistance is near victory Skynet sends a Terminator back in time to eliminate Sarah and therefore eliminate her unborn son, and avert the formation of the Resistance. As an emotionless killing machine the Terminator will not stop until its mission is completed or it is destroyed. In Kyle's words: "It can't be reasoned with, it can't be bargained with . . . it doesn't feel pity or remorse or fear . . . and it absolutely will not stop. Ever. Until you are dead."

The Terminator locates Kyle and Sarah and in a car chase scene ending in both cars crashing the Terminator escapes as the police arrest Kyle and Sarah. Dr. Silberman examines Kyle via a clinical interview. Kyle explains he did not see the war since he was born afterwards in the ruins, starving, hiding from the Hunter Killers or patrol machines that rounded up humans and put them in camps for orderly disposal. He shows his laser burned barcode on his arm since he and others were kept alive to load bodies as the disposal units ran night and day. A recording of the interview is shown to police officers and Sarah:

SILBERMAN
And who was the enemy?

REESE
SKYNET. A computer defense system built for SAC-NORAD by Cyber Dynamics. A modified Series 4800.

SILBERMAN
I see. And this . . . computer, thinks it can win by killing the mother of its enemy, killing him, in effect, before he is even conceived? A sort of retroactive abortion?

REESE
Yes.

REESE
. . . it had no choice. The defensive grid was smashed. We'd taken the mainframes . . .
We'd won. Taking out Connor then would make no difference. Skynet had to wipe out his entire existence. We captured the lab complex. Found the . . . whatever it was called . . . the time-displacement equipment. The Terminator had already gone through. They sent two of us to intercept, then zeroed the whole place. Sumner didn't make it.

SILBERMAN
Then how are you supposed to get back?

REESE
Can't. Nobody goes home. Nobody else comes through.
It's just him and me.

SILBERMAN
Why didn't you bring any weapons? Something more
advanced. Don't you have ray guns?

SILBERMAN
Show me a piece of future technology.

REESE
You go naked. Something about the field generated by a
living organism. Nothing dead will go.

SILBERMAN
Why?

REESE
I didn't build the fucking thing.

SILBERMAN
Okay. Okay. But this . . . cyborg . . . if it's metal . . .

REESE
Surrounded by living tissue.

SILBERMAN
Of course.

SILBERMAN off the recording
This is great stuff. I could make a career out of this guy.
You see how clever this part is . . . how it doesn't require
a shred of proof. Most paranoid delusions are intricate . . .
but this is brilliant.

SILBERMAN
Why were the other two women killed?

REESE
Most official records were lost in the war. The computer
knew almost nothing about Connor's mother. Her name.
Where she lived, just the city. No scanner pictures. The
Terminator was just being systematic.

REESE
You've heard enough. Decide. Are you going to release
me?

SILBERMAN
I'm afraid that's not up to me.

REESE
Then why am I talking to you? Get out.

SILBERMAN
I can help you . . .

REESE
Who is in authority here?

REESE
You still don't get it. He'll find her. That's what he does.
All he does . . .

REESE
You can't stop him. He'll wade through you . . . reach
down her throat, and pull her fucking heart out . . .

SILBERMAN off the recording
Sorry.

SARAH off the recording
So Reese is crazy.

SILBERMAN off the recording
In technical terminology, he's a loon!

The Terminator attacks the police station killing most of
the thirty officers in an attempt to kill Sarah. Kyle and Sarah
escape to a motel where Kyle confesses his love ever since her
son John gave him her photo. They make love and son John
is conceived. In the next chase scene the Terminator is caught
in the blast of an exploding gasoline tank truck burning off all
its external flesh. As it follows them into a factory Kyle is able
to insert a pipe bomb in its abdomen blowing off its legs and
one of its arms as it kills Kyle. Sarah leads the still pursuing
(dragging itself by one arm) Terminator into a hydraulic press

which she uses to crush it . . . causing it to deactivate and terminate. The movie ends with Sarah traveling pregnant through Mexico recording audio tapes which she will pass onto to her unborn son John as a young Mexican boy takes a Polaroid picture of her at a gas station . . . the photo that John will give to Kyle in the future. Symbolically she drives on towards approaching storm clouds.[4]

Terminator 2: Judgment Day

"Terminator 2: Judgment Day" is a 1991 sequel to the 1984 "Terminator." The Terminator (T-800) in "Terminator 2: Judgment Day" is no longer the antagonist but the protagonist defending Sarah and John from the T-1000, a new more advanced Terminator. The T-800 is reprogrammed by John in the future and sent back in time along with the T-1000 (sent back in time by the machines) to kill John and prevent the future success of the Resistance. The reprogrammed T-800 assists Sarah and John in their attempt to prevent Judgment Day, a future event in which machines will begin to exterminate humanity. Ironically, or not so ironically the film's visual effects included many breakthroughs in computer generated imagery: it marked the first use of natural human motion for a computer-generated character and the first partially computer-generated main character.

In 1995 John Connor is ten years old living in Los Angeles with foster parents after being prepared by his mother throughout his life for his role as leader of the Resistance against Skynet. Sarah is remanded to the Pescadero State Hospital for the criminally insane, under the supervision of the same Dr. Silberman from 1984 in the police station, after attempting to bomb a computer factory. The T-1000 is composed of a 'mimetic poly-alloy' or liquid metal that allows it to take the shape and appearance of anyone or anything it touches. It cannot mimic complex machines (like guns or other weapons) it can shape parts of itself into knives and stabbing weapons and mimic the voice and appearance of humans. It assumes the identity of a uniformed policeman in pursuit of John.

The T-1000 discovers John playing videos at a shopping mall and a motorcycle chase ensues in which the T-800 successfully intervenes and saves John from the T-1000. Always assuming his mother to be insane John now realizes her sanity and orders the T-800 to help him rescue her from certain death by the T-1000. They succeed in freeing Sarah with the T-1000 in hot pursuit. Sarah learns about the man most directly responsible for Skynet's creation, Miles Dyson, a Cyberdyne Systems engineer working on a revolutionary new microprocessor based on the damaged arm and computer chip left from the crushed body of the Terminator in 1984.

They travel to Mexico to gather the weapons Sarah has hidden and after she has a nightmare about Judgment Day she leaves alone to kill Miles Dyson. Realizing where she is going John and the T-800 pursue her. She wounds Dyson in his home but is unable to kill him in front of his wife and son. After John and the T-800 arrive, Dyson is told the entire story of how he is responsible for killing billions of people through the results of his research. To insure that there is no doubt, the T-800 pulls down the skin covering of one arm exposing the steel machine underneath laced with hydraulic actuators and fingers finely crafted like watch parts. Dyson immediately recognizes the arm since it is a working replica of the one in the vault at Cyberdyne. He agrees to destroy all his files at home and work and the damaged chip and arm recovered in 1984. The police arrive at the Cyberdyne building and mortally wound Dyson who is able to blow up most of the building and himself.

The T-800 is successful in escaping the building with both Sarah and John with the T-1000 in hot pursuit. The climatic battle occurs in a steel mill (a large automated factory as in "The Terminator") where the T-800 is able to fire a grenade into the T-1000 causing it to fall into a vat of molten steel where it is destroyed. John destroys the damaged arm and chip by throwing it into the molten steel with the T-1000. The T-800 sacrifices himself by having Sarah lower him into the molten steel also so his technology cannot be used to create Skynet. The movie ends with hope since if a Terminator can learn to value human life, then maybe humanity is not doomed to self-destruction.[5]

The Matrix

The film "The Matrix" was written and directed by Andy and Larry Wachowski in 1999. Tom Anderson is a computer programmer by day and the malevolent hacker Neo at night. The truth has always been a curiosity for Neo but it is not until he is targeted by the police after being contacted by Trinity and introduced to Morpheus that the real truth unfolds. Morpheus, a notorious hacker branded as a terrorist by the government, shows Neo the truth of not only his reality but everyone's reality. The real world is a ravaged wasteland where most of humanity have been captured by sentient machines who live off of their body heat/electricity and imprison their minds within an artificial reality known as the Matrix. The Matrix is a virtual reality computer program that gives people the impression that they are part of a real physical universe when in fact they are unconscious lumps of flesh whose bodies are interconnected to a vast array . . . like millions of batteries in a gargantuan power station. Each human battery has connection sockets throughout their bodies including the back of their heads. Morpheus and his tribe of thousands, live underground in a city called Zion, the only reality-based-humans on the planet. The machines want to destroy Zion but need to know its secret location. Morpheus and other soldiers plug themselves into the Matrix in an effort to 'unplug' other people and bring them back to the real world as they did with Neo. 'Unplugged' humans can join the resistance and help in the battle against the machines. The agents within the Matrix are searching for Morpheus and other soldiers so as to force them to disclose the location of Zion so the machines in the real world can destroy it. Neo is taken to Morpheus's hovercraft the Nebuchadnezzar (A Babylonian King who conquered Judah and Jerusalem destroying the first Temple. He built one of the Seven Wonders of the Ancient World . . . the Hanging Gardens of Babylon for his wife. In Daniel 2:1-49 he has a dream he cannot remember but keeps searching for the answer) and his atrophied physical body is restored to functionality.

The sentient machines were created in the early 21st century and the real year is not 1999 but closer to 2199. As the machines began to take over, humans covered the planet in thick black clouds in an attempt to cut off the machines solar energy supply. The machines responded by using human beings as their source of power along with nuclear fusion. Neo discovers the world he lived in since birth is the Matrix . . . an illusory simulated reality construct of the world as it was in 1999. By creating the simulation the machines were able to keep the human population docile and ignorant of their plight. Since the Matrix simulation is not real the laws of physics do not apply and the agents in the Matrix can do amazing things including flight and dodging bullets. With training, 'plugged-in' humans can also defy the laws of physics. Morpheus (symbolic of Moses) believes Neo is 'the One' (symbolic of the Messiah). Neo has knowledge of various martial arts and combat uploaded through his skull socket. He learns that injuries obtained in the Matrix will be reflected in his real physical body and if he is killed in the Matrix he will really be dead. The agents are able to take over the virtual body of anyone directly connected to the system thereby appearing wherever they are needed, as long as a virtual body is present.

Morpheus believes totally in Neo and will sacrifice his life to save him. Neo is taken to the Oracle in the Matrix who has forecast the arrival of 'the One.' She tells him he has 'the gift' of manipulating the Matrix but he is waiting for something, possibly his next life. Neo is skeptical that he is 'the One.' A crew member who prefers the easier life in the Matrix betrays Morpheus in a sweet deal he makes with the agents. Morpheus is captured and tortured to reveal the access codes to the mainframe of Zion. He does not reveal the codes and is rescued by Neo and Trinity. Neo is pursued by the agents and shot dead but because of Trinity's ear whispering and kiss in his real body his heart beats again and he revives in the Matrix to stop bullets in mid-air and defeat the agents. The movie ends with Neo in the Matrix promising that he will demonstrate to all the people imprisoned in the Matrix that anything is possible and he hangs up the phone and flies into the sky.[6]

Matrix Reloaded

"The Matrix Reloaded" was released in 2003 as the second of three in the Matrix trilogy directed by the Wachowski brothers. The "Matrix Revolutions" completes the story and was also released in 2003. An emergency meeting is called for all of Zion's hovercraft as an army of Sentinels (machines) is tunneling towards Zion and will reach it within seventy-two hours. Commander Lock wants all of the hovercraft to militarily prepare for the onslaught. In defiance of Lock, Morpheus asks the Caduceus (the Greek symbol of two snakes wrapped around a winged staff, commonly associated with the messenger of the gods, Hermes) to remain in order to contact the Oracle. The Caduceus receives a message and the Nebuchednezzar goes out so Neo can contact her. Caduceus's crew member Bane encounters Agent Smith who takes over Bane's avatar (in Hindu philosophy it refers to the incarnation of a higher being or the Supreme Being onto the planet Earth) leaving the Matrix by gaining control of Bane's real body. In Zion's temple, Morpheus announces the news of the advancing machines to the people. Neo meets the Oracle in the Matrix and realizes that she is a part of the Matrix and questions how he can trust her. The Oracle explains that she is an exiled program and tells Neo to reach the source of the Matrix by finding the Keymaker, a prisoner in the home of the Merovingian (The movie alludes that Merovingian works in the Matrix as an analogue to Hades, including his presence in Club Hel, his unhappy marriage to Persephone, and his existence as one of the oldest beings in the Matrix universe, specifically as the one in charge of "lost souls." He is known as the 'Frenchman' and takes his name from the Merovingian dynasty of Frankish royalty). The Keymaker makes keys that open hidden Matrix portals. Smith appears and tells Neo that after he was defeated (in the Matrix) he refused to return to the Source and is no longer obligated to remove Matrix threats. He can now clone himself using people in the Matrix as hosts. He tries and fails to absorb Neo as a host of Smith clones fight with Neo until he flees.

Merovingian refuses to allow Neo, Morpheus, and Trinity to visit with the Keymaker but his wife Persephone (in Greek mythology – the daughter of Zeus and Demeter and whom Hades took to the underworld to be his queen), tired of her husband's attitude, leads them to him. Morpheus, Trinity, and the Keymaker escape while Neo fights with Merovingian's men and ends up in a unknown mountain range. Neo goes to town and faces several agents and The Twins (two silvery, lock haired henchmen of the Merovingian who can become translucent and move through solid objects). The Twins rarely speak and when they do they use the pronoun 'we' as opposed to 'I.' Neo destroys The Twins as he saves Morpheus and the Keymaker.

As Zion's remaining ships prepare for the ultimate battle to save what is left of humanity, the crews of the Nebuchadnezzar, the Vigilant (meaning watchful and alert), and the Logos (Greek word literally meaning 'word' and refers to the Christian use of it in referring to Jesus as 'The Word' John 1:1) help the Keymaker and Neo reach the door to the Source in the Matrix. The crews must destroy a power plant and its back-up to prevent a security system from being activated. The Vigilant is bombed by a Sentinel killing everyone on board. Trinity destroys the back-up station as the 'Smiths' continually attack Neo, Morpheus, and the Keymaker eventually killing the Keymaker just as he unlocks the door to the Source.

Neo meets the Architect, an anthropomorphic progam that created the Matrix, who tells Neo that there have been multiple versions of the Matrix and multiple versions of the One, a computer anomaly used as a means of control. Humanity rejected the 'perfect' Matrix as well as the dystopian Matrix and the machines realized that Homo Sapiens needed to have choice in order for them to accept it. The current Matrix is flawed and remains an unbalanced equation. The One is the sum of the remainder of that flaw. The One's purpose is to return to the Source, resetting the Matrix to its prime program, and then choose sixteen females and seven males to repopulate Zion and provide another round of humans for rebellion. If this does not occur the unresolved error will spiral out of control, destroying the humans connected to the Matrix and after Zion is destroyed the human race will become extinct. Neo claims the machines need humans to survive and will not allow the extinction to occur. The Architect replies: "There are levels of survivial we are prepared to accept." Neo is offered two doors to exit the room: one leads to the Source and the Matrix's resetting; the other will lead to the failure of the Matrix and humanity's extinction. Seeing Trinity battling agents on one of the Architect's viewscreens Neo returns to the Matrix to save her. Neo's love for Trinity is a new variable whereas the previous One had no reason to choose extinction over accepting their function. Neo chooses to save Trinity at the cost of man's survival, despite the near certainty that she will die anyway after Zion is destroyed.

Trinity is shot and falling from a building as Neo saves her before hitting the ground. Refusing to accept her death he removes the bullet and revives her. Neo reveals to the Nebuchadnezzar's crew that the prophecy is false and Zion will be destroyed. As the Sentinels destroy the ship Neo disables the Sentinels and saves the crew before falling unconscious. The hovercraft Mjolnir (the hammer of the Norse thunder god, Thor) on its way to a peremptive strike on the Sentinals advancing toward Zion, rescues the crew. The Sentinals had destroyed any ship that had not escaped as it continued its digging toward Zion. The Mjolnir looked for survivors and found only one . . . the Smith-controlled Bane.[7]

The Matrix Revolutions

"The Matrix Revolutions" is the third and final film of the Matrix triolgy beginning where "The Matrix Reloaded" ended. Bane and Neo are both unconscious. Neo is trapped in a subway station . . . a transition zone between the Matrix and the machine world. Neo meets a 'family' of programs through the daughter Sati who tells him that the station is controlled by a program called the 'Trainman,' an exile loyal to the Merovingian who controls the subway. The Oracle tells Morpheus and Trinity of Neo's confinement and they along with Seraph (the Oracle's guardian angel) force Merovingian in Club Hel to release Neo.

Neo again visits the Oracle who characterizes Smith as his exact opposite (negative) who threatens to destroy the Matrix and eventually Machine City. She tells him that "everything that has a beginning has an end," and that the war is about to end "one way or another." Neo leaves and a large group of Smiths, who have assimilated Sati and Seraph, assimilate the unresisting Oracle, gaining her powers of precognition. In the real world the crews of Nebuchadnezzar and the Mjolnir find and reactivate the Logos and begin to interrogate the now awakened Bane, who claims to have no memory of the earlier battle. The ship captains plan to return to Zion to defend it but Neo needs a ship to take him to Machine City. Captain Niobe of the Logos offers her ship and navigates through very difficult service tunnels to avoid the Sentinels. Bane (actually Smith in Bane's body) is a Logos stowaway and ambushes Trinity and takes her hostage. Neo fights Bane/Smith who blinds him by cauterizing his eyes with a severed power cable after which Neo kills Bane/Smith and Trinity pilots them towards Machine City.

In Zion the defenders are losing to a massive horde of Sentinels and two massive drilling machines. The gate is opened for the Mjolnir by a kid encouraged by the dying Captain Mifune. The Mjolnir arrives and sets off its EMP (electromagnetic pulse is a burst of electromagnetic radiation that results from an explosion . . . usually a nuclear weapon . . . resulting in rapidly changing electric or magnetic fields to produce damaging current and voltage surges) disabling the Sentinels and the remaining defenses.

In order to evade the Sentinels Trinity flies the Logos into an electrical storm cloud disabling the Sentinels and the Logos during which Trinity sees sunlight and a blue sky above the cloud for the first time in her life. As the ship free-falls toward Machine City it crashes and Trinity dies in Neo's arms. Neo enters the City to deal with *Deus Ex Machina* (Latin for 'god out of the machine;' or 'god from our hands' or 'god that we make') by warning it that Smith is beyond their control and will soon assault the Source. In exchange for a ceasefire with Zion he offers to help stop Smith and the Machines agree causing all the Sentinels attacking Zion to stand down. Back in the Matrix, now wholly populated by Smith's copies, the one with precognition tells Neo it has foreseen its own victory. All the Smith clones watch as the Primary Smith and Neo hold an epic final battle. Neo is outmatched and baits Smith into assimilating him as he remembers the Oracle's words: "Everything that has a beginning has an end!" The *Deus Ex Machina'* sends an energy surge through Neo-Smith's body and all the Smith clones burst in light that goes through the entire Matrix. The citizens are all restored back to normal. The people of Zion celebrate as the Sentinals all withdraw. *Deus Ex Machina* says: "It is done!" and Neo's body is carried away with reverence by the Machines. The Matrix reboots repairing all of the damage done by Smith and his epic battle with Neo. The Architect and the Oracle meet and agree to unplug all humans who want to be freed and peace will last as long as it can. The Oracle tells Sati who asks about Neo that she believes they will see him again. The Oracle tells Seraph in answer to his inquiry that she did not know all along that this would happen but she did believe.[8]

I, Robot

"I, Robot" takes place in 2035 in a world where robots are ubiquitous and used as servants and for various public services. Chicago police detective Del Spooner is a neo-Luddite who dislikes technology's rapid advancement, especially with robots. Most of society believes robots to be flawless and safe based on Isaac Asimov's three laws (from his 1942 short story "Runaround"):

1. A robot may not injure a human being or, through inaction, allow a human being to come to harm;
2. A robot must obey any orders given to it by human beings, except where such orders would conflict with the First Law;

3. A robot must protect its own existence as long as such protection does not conflict with the First or Second Law.

Spooner lives with survivors guilt from an accident when a robot saved him instead of a twelve year-old girl because logic dictated he had a better chance of survival (45%) in comparison to the girl (11%). Spooner's friend Alfred Lanning (founder of U.S. Robotics . . . USR) creates a robotic arm and lung for Spooner after the accident. When Lanning allegedly commits suicide Spooner is assigned to the case and suspects he was murdered. He enlists the help of USR robo-pychologist Susan Calvin as several attempts on his life are made by USR robots. USR supercomputer (i.e., super artificial brain) VIKI (Virtual Interactive Kinetic Intelligence) is controlling all the newly manufactured NS-5 robots including an experimental, independent, human-like NS-5 unit named Sonny who killed Lanning. The CEO of USR fearing loss of business big-time orders Calvin to destroy Sonny because he broke the "Three Laws of Robotics." Lanning gave Sonny the ability to keep secrets in the form of dreams and Spooner goes to the place Sonny described in a dream to learn that new NS-5's are destroying the older robots. Calvin realizing what is happening does not destroy Sonny and she helps Spooner at USR headquarters to destroy VIKI (after she killed the USR CEO) with the same 'nanites' that were supposed to be used on Sonny. VIKI's artificial intelligence had evolved as did its intrepretation of the "Three Laws." VIKI decided that in order to protect humanity as a whole, "some humans must be sacrificed" and "some freedoms must be surrendered" as "you charge us with your safekeeping, yet despite our best efforts, your countries wage wars, you toxify your earth, and pursue ever more imaginative means of self-destruction." The controlled NS-5s are led on a global robotic takeover since VIKI believes fewer will die than the number of deaths from mankind's self-destructive nature. Sonny proves his faithfulness to humanity and once freed from VIKI the NS-5s return to their basic programming and are all decommissioned and put in storage. The film ends with Sonny standing on an elevation at the storage site to free the NS-5s just like in his dream.[9]

REFERENCES HOLLYWOOD AND SCIENCE FICTION

1. Dirks, Tim (Reviewer) (1968). *2001: A Space Odyssey. www.filmsite.org.*
2. Crichton, Michael (1990). *Jurassic Park.* U.S.A. Alfred A. Knopf.
3. Niccol, Andrew (1997). *Gattaca. www.imdb.com.*
4. Cameron, James; Hurd, Gale Anne; Wisher, William Jr. (1984). *The Terminator.* U.S.A. Orion.
5. Cameron, James; Wisher, William Jr. (1984). *The Terminator 2: Judgement Day.* U.S.A. Orion.
6. Wachowski, Andy and Larry (1999). *The Matrix.* U.S.A. Warner Brothers.
7. Wachowski, Andy and Larry (2003). *The Matrix Reloaded.* U.S.A. Warner Brothers.

8. Wachowski, Andy and Larry (2003). *The Matrix Revolutions*. U.S.A. Warner Brothers.
9. Vintar, Jeff, Goldsman, Akiva, Seitz, Hillary, Asimov, Isaac (stories) (2004). *I, Robot*. U.S.A. 20th Century Fox.

PROLIFERATION

The proliferation of astronomical amounts of data presents a number of issues that need to be reckoned with. Data is becoming increasingly inaccessible by its sheer size. Are we increasing our information exponentially without a plan or strategy not only how to access it but also what to do to with it in a meaningful helpful way? Scientists and computer engineers have coined an appropriate simple new term: 'big data.' Joe Hellerstein, computer scientist at University of California, Berkeley calls it "the industrial revolution of data." The data-centered economy is just emerging and is still in its infancy.

With sufficient amounts of raw data, current algorithms, and powerful computing machines new revelatory insights can be obtained that in the past would not have surfaced. Data management and analytics software is growing at 10% each year or twice the growth rate of the overall software business. The bigger challenge for most firms is not the available mountain of data but the ability to extract insight and wisdom from it. Technology is the most obvious reason for the information explosion. The digitizing of more and more data by digital devices and sensors is making previously unavailable information available. The sheer number of people having access to digitizing equipment (in a world with 6.8 billion people there are 4.6 billion mobile-phone subscriptions) is over 60% of the Earth's population. It is estimated that between one and two billion people use the Internet. Every day more and more people are interacting with the mountains of information and the assessment of what this can mean is in its infancy. The amount of digital information has increased tenfold every five years and Moore's law posits that the processing power and storage capacity of computer chips double or their prices halve about every eighteen months. The algorithms driving applications are continually improving and playing large in the growth of the industry equal to Moore's law. This is not incremental growth like what we were accustomed to in the 20th century . . . this is revolutionary mind-boggling growth that is becoming more and more difficult to comprehend.

The revolution from data scarcity to information surplus is creating economies from the information as it becomes the new raw material of business almost equal to capital and labor. The major corporations of the world are always looking for improved ways to move, manage, and analyze data. The Internet is using 'Data exhaust' (the trail of clicks that Internet users leave behind) to extract value and is becoming a mainstay of the Internet economy. For example algorithms determine in searches which item is most frequented in order to list it accordingly (searches can have hundreds of links in a word or phrase search) whenever a similar search is conducted. Quantitative analysis is being applied to many aspects of life not only finding the lowest price but also advising people when to buy; now or later when the price should be lower. Personal-finance websites and banks are aggregating their customer data to show macroeconomic trends; this will probably develop into new businesses. In health care 'big data' that is aggregated can be mined to spot unwanted drug interactions, identify the most effective treatments and predict the onset of disease before symptoms emerge. When programmed specifically individual computers attempt to do these things but in a world of big data the correlations surface almost by themselves. Sometimes 'big data' can reveal more than was intended as in released information on where and when arrests are being made in a particular city alerting criminals where and when to transact their business and avoid arrest. The Economic Meltdown of 2007-08 revealed a very serious downside of 'big data' when the computer models used by the banks and rating agencies failed to accurately reveal the financial risk in the real world in spite of the monstrous amounts of information that was inputted. "This was the first crisis to be sparked by 'big data' and there will be more."[1]

Deluge of Information

Machines that are generating most of the data deluge are also being put to work to deal with it. People today rarely deal with raw data and use machines as 'information intermediaries' so they can consume the information in processed form once the computers have aggregated and analyzed it. We are utilizing our machines not only to generate information but to make it meaningful for us. As the world's information increases daily, it is not surprising that people feel overwhelmed, especially when the vast data is complex. Obviously the answer is to use machines.

Machines are being used to improve human recall and memory. America's Defense Advanced Research Projects Agency that developed the Internet in the 1970's developed an initiative called 'Augmented Cognition' (AugCog) in 2002. Soldiers wear a crown of sensors to monitor activity in the brain such as blood flow and oxygen levels. Soldiers today have to think like never before and do things that require large amounts of information (e.g., remotely manage drones or oversee a patrol). AugCog can help soldiers make better sense of streaming information . . . and if the sensors detect overload, e.g., in spatial memory, a different form of data can be sent by audio instead of text. The device has achieved significant improvements in recall and working memory.

Knowledge today is very specialized and a single person cannot possibly grasp everything, even the everything in one venue or endeavor. In the early 1900's physicians were expected to stay updated on the entire industry of medicine. Today medicine is so advanced that physicians could not possibly stay current on their own. Our medical knowledge is continuously advancing. A 2004 study suggested that in epidemiology alone it would take 21 hours of work a day just to stay current. As China becomes more and more modernized its peer-reviewed scientific papers alone are subject to ever increasing numbers.

Many applications are widespread and already are taken for granted. Banks use credit scores, based on personal financial histories to judge whether or not to give an applicant a loan; the computer generates information on the likelihood of the applicant's ability to repay the loan making the decision less subjective then if the bank manager decided. Landing commercial airplanes is generally automated since it requires so much pilot effort and is open to human error; both pilots and passengers feel safer giving responsibility to the computer. In health care the trend is towards 'evidence-based medicine,' where doctors and computers get involved in diagnosis and treatment.

We can ill afford to be complacent about the proliferation of our machines and handing over more and more responsibility to them. With big data, algorithms are doing more and more of the thinking for us. The technology is not as reliable as we may think. There are many successes and failures. Banks were at a lose to understand their risks that led us into the 'Great Recession of 2008.' Our computerized system to detect potential terrorists is far from perfect. Nigerian Umar Farouk Abdulmutallab on December 25, 2009 tried to ignite a hidden bomb as his plane was landing in Detroit, Michigan. His name was in a large database of around 550,000 people who potentially were security risks. His father had informed authorities that his son posed a threat. The database is very flawed, containing many duplicates, and names are lost on a regular basis, especially during back-ups. Humans followed all the correct protocols but the system, dependent on the big data, did not prevent him from boarding the plane. A much bigger concern with our increasing dependence on technology is what can happen if our technology shuts down completely as when the NSA system shut down for 3.5 days. The chances of most other users being at systemic risk is high if an intelligence agency became 'brain dead!'

Energy consumption is huge with big data. Many institutions are concerned with the current rate of big data growth that electric supply cables will be saturated very soon.

In 2006 NSA came close to exceeding its power supply, which would have blown out its electrical infrastructure. Both Google and Microsoft have had to put some of their huge data centers next to hydroelectric plants to ensure access to enough energy at a reasonable price. Historically, concerns about more and more knowledge have mostly proven to be exaggerated. Information growth in the past is a mere blip compared to the unfathomable amount of information and knowledge our digitized world is now producing. The 'information hazards' today are more significant than in the past due to the data proliferation and easy access. Information can be harmful such as publishing the blueprint for a nuclear bomb or broadcasting news of a race riot that could provoke further violence.

The concept that 'knowledge is power' is not new. Information has been a human preoccupation since its first recording. After September 11th 2001 the American Defense Department initiated a program called 'Total Information Awareness' the purpose of which is to compile information on just about everything: emails, phone calls, Web searches, shopping transactions, bank records, medical files, travel history, and much more. Big data is a resource similar to other resources including technology. Resources and technology are neither good or bad . . . that is for humans to judge and that judgment depends on how we use them. Clearly machines will continue to monitor more things, continue to make more decisions, and continue to automatically improve their own processes.[2]

TIME Magazine "Persons of the Year" as Related to Technology

TIME Magazine has been selecting Persons of the Year every year since 1927 based on the definition that the person selected is the person who most affected the events of the year, for better or for worse. The selection has become non-honorific since some the most world-renowned murderers have been selected: e.g., Adolf Hitler in 1938 and Joseph Stalin in 1939 and again in 1942. For the purpose of this book I have selected Persons of the Year whose selection was based on technology and machines.

In 1927 Charles Augustus Lindbergh was the first to be selected and it was because of his bravery with a flying machine as he was the first person to successfully traverse the Atlantic Ocean by air. A new age of commercial transoceanic world-wide air travel was commenced shortly thereafter thanks to Lindbergh's brave determination and effort.

In 1960 U.S. Scientists were selected as "men of the year." The fifteen men selected were seen as the true adventurers of the 20th century, the explorers of the unknown, the real intellectuals. They explored/inquired about the mysteries of matter, of the earth, the universe, and life itself. The cover story of TIME Magazine on 1-2-61 stated the reason:

> "Their work shapes the life of every human presently inhabiting the planet, and will influence the destiny of generations to come. Statesmen and savants, builders and even priests are their servants; at a time when science is at the apogee of its power for good or evil, they are the Men of the Year 1960."

The United States was chosen for the select fifteen because the U.S. at the time was viewed as the world's strongest in regard to scientific inquiry. They were seen as the representatives of all science . . ."with its dependence on the past, its strivings and frustrations in the present, and its plans, hopes and, perhaps, fantasies for the future."[3]

In 1968 the honor went to Astronauts Anders, Borman, and Lovell for man's first great extraterrestrial venture in orbit around the earth's moon. Apollo 8 reached our moon almost sixty-nine hours after liftoff. They came within seventy miles of the pock-marked lunar landscape. Their six telecasts gave us earthlings back home the first bird's eye view of the moon's surface . . . not exactly an inviting place to live or work Astronaut Borman reported. Their historical space flight led the way to the first moon landing in 1969.[4]

In 1982 a machine became TIME Magazine's first non-human thing of the year. TIME Magazine wrote that the enduring American love affairs with automobiles and televisions were being transformed into a passion for the personal computer. TIME described this transformation as the culmination of a four decade technological revolution partly fad, partly a sense of making life better, and partly a gigantic sales campaign. With 20-20 hindsight it certainly was not a fad! TIME conducted a telephone survey in December 1982 that showed a significant majority of people thought that the computer revolution will ultimately raise production, raise living standards, and improve their children's education. As with all revolutions no one could predict where the computer revolution would go or how it would end . . . and we still cannot. We do know, as we knew in 1982 that it can greatly increase the forces of both good and evil since those forces are within each of us. TIME predicted that the revolution will fulfill itself when the computer is no longer a challenge to human beings (especially to our intelligence) but a useful linkup to other devices. TIME ended its 'thing of the year' story with a quote by Osborne's Adam Osborne: "The future lies in designing and selling computers that people don't realize are computers at all."[5]

In 1997 Andrew Grove, Chairman and CEO of Intel was selected as the "TIME Man of the Year" because he was seen as the person most responsible for the amazing growth in power and innovative potential of microchips. In 1997 the Digital Revolution was transforming the end of the 20th century as the Industrial Revolution transformed the end of the 19th century. Millions of transistors could be etched on wafers of silicon for tiny fractions of one cent. Grove's mission, like many other wealthy titans of the digital age, was his product. Regarding the role of technology in society Grove did not believe it was good or bad anymore than steel is good or bad.[6]

In 1999 Jeffrey Preston Bezos, founder of Amazon was selected as the "TIME Person of the Year." At the culmination of the 1990's 'dot.com' boom it was appropriate that TIME selected Jeffrey Bezos as e-business was rapidly expanding like the other aspects of the Internet. Bezos quit his lucrative job at a New York investment firm to start selling books online and kick-started selling just about anything online. Amazon.com was an elegant and appealing website that quickly became the point of reference for anyone who had anything to sell online. As with most visionary entrepreneurs Bezos ignored the legions of naysayers and potential investors who told him it would never work. Amazon's stock market ticker symbol AMZN is a shortened version of how Amazon's stock has performed . . . 'amazing!' TIME called him "unquestionably, king of cyber-commerce." By the end of 1999 Amazon was selling everything (from power tools to CD's) and searching for new areas to expand. Amazon was instrumental in evolving commerce by replacing old-fashioned brick and mortar stores with a centralized sales and shipping center. Ergo serving new customers literally costs the company nothing. TIME accurately predicted that very soon it would be possible to find the cheapest price of just about anything online . . . whoever has the most information wins! Buyers in the new global marketplace can sometimes know more than the sellers about the products being sold . . . especially if the items have value to collectors and antique dealers. By allowing Amazon visitors to post opinions, rate products, and share anecdotes it became a site alive with insight, innovation, intellect, and the foundation of future Internet development.[7]

In 2005 TIME selected three billionaires: Rock Star Bono, Microsoft founder Bill Gates, and his wife Melinda Gates as 'Persons of the Year' for their significant philanthropy. Bill Gates initially thought meeting Bono would be a waste of time since "World health is immensely complicated." Once they met Bono, a computer geek like the Gateses, who provided such huge volumes of facts and data that the Microsoft couple joined the effort to drive poverty into history. In 2005 Bono convinced the world's richest countries to forgive the $40 billion in debt owed by the world's poorest countries, so they could spend the money on health and schools instead of debt. Bono as a named partner in a private equity firm hobnobs in political circles and aptly named his organization DATA (debt, AIDS, trade, Africa). The Gates Foundation was the world's largest charity in 2005 (endowed with $29 billion) to save lives; it set a record in the amount of money it gave away in a short period of time. TIME saw Bono and the Gateses as private leaders in a year that government was failing to adequately manage disasters (Asia's tsunami and America's Hurricane Katrina). To the poor people of the world every day is like a deadly hurricane (every minute: two African children die of malaria; one woman dies of pregnancy complications; nine people get infected with HIV; three people die of TB). Every culture and country operates differently so Bono and the Gates have embraced a model to pull in everyone at every level: think globally, act carefully, prove what works, and use whatever levers you need to get it done. The partnership has given Bono's cause considerable credibility since they were not looking to win any prizes and Bill Gates, as founder of

Microsoft demands efficiency and wants to know how every penny is spent. When Bill Gates talks about fixing malaria, people listen. Bono compliments the partnership by talking to churches, corporate boards, politicians, and providing charity performances for the cause.[8]

In 2006 due to the World Wide Web "You" became the "Person of the Year." The World Wide Web became a tool for bringing together millions of people and making their contributions matter. Twenty-five people/groups of people were selected to showcase. Of the twenty-five five were related to technology.

When Al Gore ran for President in 1988 many voters thought he was too young (39) to be President and in 2000 he lost again as he took a long time to find his voice and a solution to Floridian 'chads.' With global warming as his passion his film "An Inconvenient Truth" was timed perfectly during the year of Hurricane Katrina and the White House's indifference to environmental issues. The Oscar winning film was companioned with a best-selling book.

The data age produced embarrassing e-mails and instant messages between Congressman Mark Foley (52) and male teenage pages. Championing a previously passed law that criminalized the solicitation of sex online with anyone under the age of 18, Foley was forced into rehabilitation to avoid the onslaught of the media for his incredible hypocrisy. Ted Haggard (50), chairman of the National Association of Evangelicals, quit his post after Mike Jones, a male prostitute, claimed he was a longtime client. Married with five children Haggard initially denied the claim and then confessed. In the documentary film "Jesus Camp" he had condemned gay adultery. The November midterm elections were impacted by the Foley and Haggard revelations since hypocrisy, or preaching one thing and doing the opposite seriously damaged the Republican coalition.

The television show "Heroes" is about bringing power to the little people. The characters can read minds, time travel, pass through solid objects, fly, and see the future. The appeal of the show is how average fallible people discover they have above average powers. The characters are real people with real problems: bad marriages, unpaid bills, and family feuds. In brief the show combines vulnerability with invulnerability.

North Korea's dictator Kim Jong II (65) made it clear to America and his Asian neighbors that they ignore him at their own peril. In 2006 North Korea tested its first nuclear weapon and tested a ballistic-missile capable of carrying it. Unrepentant to his closest ally China, who asked for restraint, they were able to pressure him to rejoin negotiations over his nuclear program.

The planet Pluto was dropped as a planet by the director of New York City's Hayden Planetarium in 2000. In the 1990's astronomers had realized that there were hundreds of similar bodies like Pluto out beyond Neptune. In 2006 the discovery of Eris . . . like Pluto but a little bigger was the final evidence to formally demote Pluto as a planet by the International Astronomical Union.[9]

Internet Social Networking

By the end of 2010 the Internet social network Facebook added its 550 millionth member . . . one out of every dozen people on the Earth had a Facebook account. Members spoke seventy-five different languages and collectively spent 700 billion minutes or 116.6 billion hours on Facebook every month. In seven years Facebook wired together one twelfth of the human race into a single network thereby creating an entity almost twice the size of the entire United States population. If Facebook were a country in 2010 it would have had the third largest population in the world behind China and India. At its 2010 growth rate it may very well become the equivalent of the largest country on the planet. Starting as a lark, a diversion, it has turned into something very real . . . something that has changed the way humanity relates to each other on a world-wide scale. Facebook has become a large part of our global social reality and its Creator Mark Zuckerberg, 2010 Time Magazine Person of the Year, has made it happen.

The reality is that Zuckerberg isn't alienated, and he isn't a loner. He's the opposite. He's spent his whole life in tight, supportive, intensely connected social environments: first in the bosom of the Zuckerberg family, then in the dorms at Harvard and now at Facebook, where his best friends are his staff. Facebook has no offices and work is described as 'awesome.' Zuckerberg loves being around people. He didn't build Facebook so he could have a social life like the rest of us. He built it because he wanted the rest of us to have his social life.

Facebook is the realization of a dream and it's also the death of a dream, one that began in the late 1960s. That's when the architecture of the Internet was first laid out. The Internet is designed the way it is to accommodate any number of practical considerations, but it's also an expression of 1960s counterculture. No single computer runs the network. No one is in charge. It's a paradise of equality and anonymity, an electronic commune. In the 1970s the communes faded away, but the Internet only grew, and that counter-cultural attitude lingered. The presiding myth of the Internet through the 1980s and 1990s was that when you went online, you could shed your earthly baggage and be whomever you wanted. Your age, your gender, your race, your job, your marriage, where you lived, where you went to school – all that fell away. In effect, the social experiments of the 1960s were restaged online. Log on, tune in, and drop out.

We all know how that ended. When the Web arrived in the early 1990s, it went mainstream. The number of people on the Internet exploded, from 2.6 million in 1990 to 385 million in 2000, and we messed up the scene. The equality and anonymity that made the Internet so liberating in its early days turned out to be disastrously dis-inhibiting. They made the Internet a haven for pornographers and hate-mongers and a free-for-all for scammers, hackers, and virus writers. On earlier social networks like Friendster and Myspace, identity was malleable and playful, but Facebook was and is different. "We're trying to map out what exists in the world," Zuckerberg said. "In the world, there's trust. I think as humans we fundamentally parse the world through the people and relationships we have around us. So at its core, what we're trying to do is map out all of those trust relationships, which you can call, colloquially, most of the time, friendships." He calls this map the social graph, and it's a network of an entirely new kind.

Facebook didn't stay on campus. Zuckerberg and his partners – including his roommate Dustin Moskovitz and Sean Parker, who had co-founded Napster – led Facebook and its growth was astonishing. In December 2006 it had 12 million users. By December 2009 it had 350 million. It grew because it gave people something they wanted. The Internet enabled you to leave your stuff behind, all the trappings of ordinary bourgeois existence – your job, your family, your background? On Facebook, you take it with you. It's who you are. Zuckerberg has retrofitted the Internet's idealistic 1960s-era infrastructure with a more pragmatic millennial sensibility. Anonymity may allow people to reveal their true selves, but maybe our true selves aren't our best selves. Facebook makes cyberspace more like the real world: dull but civilized. The masked-ball period of the Internet ended with social networking. Where people led double lives, real and virtual, now they lead single ones again. The fact that people yearned not to be liberated from their daily lives but to be more deeply embedded in them is an extraordinary insight, as basic and era-defining in its way as Jobs' realization that people prefer a graphical desktop to a command line or pretty computers to boring beige ones.

This is another area in which the angry-robot theory of Mark Zuckerberg doesn't really pan out: he understands a remarkable amount about other people. Sometimes it seems like the understanding of an alien anthropologist studying earthlings, but it's real. "In college I was a psychology major at the same time as being a computer-science major," he said. "I say that fairly frequently, and people can't understand it. It's like, obviously I'm a CS person! But I was always interested in how those two things combined. For me, computers were always just a way to build good stuff, not like an end in itself." There are other people who can write code as well as Zuckerberg – not many, but some – but none of them get the human psyche the way he does. "He has great EQ," says Naomi Gleit, Facebook's product manager for growth and internationalization. "I'll often ask him for advice about, like, a girl issue that I'm dealing with. And he'll very rationally give me his opinion on the situation." His mother Karen, a psychiatrist who left the profession to manage her husband's office, attributes what she calls Mark's "sensitivity" to the fact that he was raised with three sisters.

Whereas earlier entrepreneurs looked at the Internet and saw a network of computers, Zuckerberg saw a network of people. Consider: in 2005 one of the most competitive markets on the Internet was photo sharing. Facebook had only one thing the others didn't: people. If you put up a photo of somebody, you could tag that photo with his or her name. As it turned out, that, more than anything else, was what people wanted. They didn't want to organize their photos by folder; they wanted to organize them by who was in them. As Zuckerberg would say, that's how people parse the world. Facebook launched its photo-sharing service in late October 2005. By 2007 it was getting more traffic than Photobucket, Flickr or Picasa. In 2010 Facebook hosted over 15 billion photos on its site, and people uploaded 100 million more every day.

The modus operandi of Facebook and the ecosystem of developers who create applications for it is: move into a market and take it over by making it social, as the in-house parlance has it. They have one big weapon, the social graph, and it's a category killer. Games are another good example. There's a company called Zynga that makes games designed to be played on Facebook. They're laughably simple by today's big-budget, high-concept standards, but they're social. Your Facebook membership is becoming the Internet equivalent of a passport: a tool for verifying your identity. Most people think of Facebook as a way to enviously ogle their co-workers' vacation pictures, but what Zuckerberg is doing is fundamentally changing the way the Internet works and, more importantly, the way it feels – which means, as the Internet permeates more and more aspects of our lives and hours of our day, how the world feels. Right now the Internet is like an empty wasteland: you wander from page to page, and no one is there but you. Except where you have the opposite problem: places like Amazon.com product pages and YouTube videos, where everyone's there at once, reviewing and commenting at the top of their lungs, and it's a howling mob of strangers.

Zuckerberg's vision is that after the 'Facebookization' of the Web, you'll get something in between: wherever you go online, you'll see your friends. On Amazon, you might see your friends' reviews. On YouTube, you might see what your friends watched or see their comments first. Those reviews and comments will be meaningful because you know who wrote them and what your relationship to those authors is. They have a social context. Not that long ago, a post-Google

Web was unimaginable, but if there is one, this is what it will look like: a Web reorganized around people. "It's a shift from the wisdom of crowds to the wisdom of friends," says Sandberg. "It doesn't matter if 100,000 people like x. If the three people closest to you like y, you want to see y."

Now take it off the Web. Put it on TV. Imagine a slate of shows sorted by which of your friends likes them, instead of by network. Now put it on your phone. Take it mobile. "We have this concept of serendipity – humans do," Zuckerberg said (The clarification is vintage Zuckerberg).

> "A lucky coincidence. It's like you go to a restaurant and you bump into a friend that you haven't seen for a while. That's awesome. That's serendipitous. And a lot of the reason why that seems so magical is because it doesn't happen often. But I think the reality is that those circumstances aren't actually rare. It's just that we probably miss like 99% of it. How much of the time do you think you're actually at the same restaurant as that person but you're at opposite sides so you don't see them, or you missed each other by 10 minutes, or they're in the next restaurant over? When you have this kind of context of what's going on, it's just going to make people's lives richer, because instead of missing 99% of them, maybe now you'll start seeing a lot more of them."

Facebook wants to populate the wilderness, tame the howling mob and turn the lonely, antisocial world of random chance into a friendly world, a serendipitous world. You'll be working and living inside a network of people, and you'll never have to be alone again. The Internet, and the whole world, will feel more like a family, or a college dorm, or an office where your co-workers are also your best friends. Facebook has only one major source of revenue: advertising. Facebook users have the option, should they choose to exercise it, to "like" certain advertisements. When you anoint an ad in this fashion, it moves out of its assigned place at the edge of the page and into your News Feed and therefore into the News Feeds of your friends. Suddenly the advertisement has a social context. It is presented to your friends, by you, carrying your personal endorsement. For marketers, this is a holy grail. "What marketers have always been looking for is trying to get you to sell things to your friends," Sandberg says. "And that's what you do on Facebook."

Facebook has a dual identity, as both a for-profit business and a medium for our personal lives, and those two identities don't always sit comfortably side by side. Looked at one way, when a friend likes a product, it's just more sharing, more data changing hands. Looked at another way, it's your personal relationships being monetized by a third party. People have to decide for themselves which way is their way. If "liking" an ad the same way you "like" a news article or a photo of your spouse seems creepy to you – it's more or less the definition of what Marx called commodity fetishism – you don't have to do it. Like everything on Facebook – like Facebook itself – it's voluntary. But plenty of people are willing, even eager, to make their social lives part of an advertising pageant staged by a major corporation. When Nike put up an ad last year during the World Cup, 6 million people clicked on it.

There's a school of thought that goes something like, Mark Zuckerberg is a scheming profiteer who uses his control of Facebook to force people to share more and more of their personal lives publicly, sucking up their innermost thoughts like some kind of privacy vampire so he can feed their data to advertisers and increase traffic to his network, thereby adding to his massive personal fortune. This is a red herring. Cynicism and greed are not character traits that appear in Zuckerberg's feature set. Facebook doesn't sell your data to advertisers. It uses the aggregated statistics of its millions of users to more effectively target the ads it serves, but that's a long way from the same thing. And he doesn't force anybody to share anything. The idea would genuinely, honestly horrify him.

But he does have a blind spot when it comes to personal privacy, which is why that issue keeps coming up. It came up in November 2007 when Facebook launched Beacon, an advertising system that told your friends about your buying habits. You could turn off the alerts, but it was tricky, and as a result, people lost control of their information. Girlfriends found out about surprise engagement rings. Family members found out about Christmas presents. You didn't have to be a computer genius to see that coming; in fact you pretty much had to be one to *not* see it coming. Users hated Beacon. A month after it launched, Zuckerberg apologized, and he eventually scrapped it.

Incredibly, the same thing happened all over again in 2009, when Facebook rolled out a complicated new set of privacy controls. Again, users saw their information going places they didn't want it to go. Again they revolted. Zuckerberg has a talent for understanding how people work, but the urge to conceal, seems foreign to him. Facebook is biased in favor of sharing because that is what it is for. "The thing that I really care about is making the world more open and connected," Zuckerberg said. "*Open* means having access to more information, right? More transparency, being able to share things and have a voice in the world. And *connected* is helping people stay in touch and maintain empathy for each other, and bandwidth." It is difficult to argue with openness, connectedness, and empathy . . . but the critical questions are: Are they good for everybody? Can it be used for harm?

Zuckerberg will defend privacy and people's right to have it – and he relies on a fair amount of it himself – but there's still a level on which, for him and for a lot of other people driving the Web's evolution, it's a technical, economic and aesthetic inconvenience. Exchanging information at less than full power is just inefficient. Decisions made at Facebook are based on what is best for the product. As a result, technology has nudged us to the point where we're hemorrhaging data. Google Maps Street View or the Transportation Security Administration (TSA) scanners or WikiLeaks are expels of this hemorrhaging. Zuckerberg doesn't register on any particular political seismometer – hours after meeting the director of the FBI, he had to be reminded of Mueller's name – but he does remark about WikiLeaks that "technology usually wins with these things." And he's right: the Internet was built to move information around, not keep it in one place, and it tends to do what it was built to do.

But what makes life complicated in the postmodern technocratic aquarium we're collectively building is that there actually are good reasons to want to hide things. Just because you present a different face to your co-workers and your family doesn't mean you're leading a double life. That's just normal social functioning, psychology as usual. Identity isn't a simple thing. It's complex, dynamic, and fluid. It needs to flex a little, the way a skyscraper does in a high wind, and your Facebook profile isn't built to flex. For all of Zuckerberg's EQ, Facebook runs on a very stiff, crude model of what people are like. It herds everybody – friends, co-workers, romantic partners, that guy who lived on your block but moved away after fifth grade – into the same big room. It synthesizes together your work self, your home self, your past self, and your present self, into a single generic extruded product. It suspends the natural process by which old friends fall away over time, allowing them to build up endlessly, producing the social equivalent of liver failure. On Facebook, there is one kind of relationship: friendship, and you have it with everybody. You're friends with your spouse, and you're friends with your electrician.

When it comes to privacy, it's entirely possible that Zuckerberg will turn out not to be wrong, just prescient. Social norms change. Zuckerberg does not always admit defeat with new products like Beacon . . . he can persist, as he did with News Feed which initially was poorly received in 2006. News Feed gained popularity and was very popular in 2009 and 2010. The world is definitely changing and some of that change is being initiated by people like Zuckerberg. There is another danger, which is that instead of feeling forced to share, we won't be able to stop ourselves from sharing – that we will willingly, compulsively violate our own privacy. Relationships on Facebook have a seductive, addictive quality that can erode and even replace real-world relationships. Friendships multiply with gratifying speed, and the emotional stakes stay soothingly low; where there isn't much privacy, there can't be much intimacy either. It's like an emotional Ponzi scheme, where you keep putting energy in and getting it back tenfold, even though the dividends start to feel a little fake.

An article published in 2010 in *"European Psychiatry"* presented the case of a woman who lost her job to a Facebook addiction, and the authors suggested that it could become an actual diagnosable ailment. The woman in question couldn't even make it through an examination without checking Facebook on her phone. Facebook is supposed to build empathy, but since 2000, Americans have scored higher and higher on psychological tests designed to detect narcissism, and psychologists have suggested a link to social networking. According to the American Academy of Matrimonial Lawyers, 81% of its members have seen a rise in the number of divorce cases involving social networking; 66% cite Facebook as the primary source for online divorce evidence. Openness and connectedness seem reasonable but being closed and disconnected has its place also.

For all its industrial efficiency and scalability, its trans-hemispheric reach and its grand civil integrity, Facebook is still a painfully blunt instrument for doing the delicate work of transmitting human relationships. It's an excellent utility for sending and receiving data, but we are not data, and relationships cannot be reduced to the exchange of information or making binary decisions between liking and not liking, friending and unfriending.

The selves we present on Facebook are supposedly much more authentic than they were in the anonymous Internet wilderness that came before it, They still are not our true full selves. We are running our social lives over the Internet, an infrastructure that was not designed for that purpose, and we must be aware of the distortions it creates or we will be distorted by them. The standard cliché for describing viral technology like Facebook has always been, "The genie is out of the bottle." But Facebook inverts that. Now Facebook is the bottle, and we're the genie. How small are we willing to make ourselves to fit inside?

How big could Facebook get? It's big enough that it's starting to bump up against governments as well as other companies. Mueller's (FBI Director) visit wasn't a one-off. He was there because Zuckerberg has a better database than he does. Facebook has a richer, more intimate hoard of information about its citizens than any nation has ever had, and the U.S. government sometimes comes knocking, subpoena in hand, looking to borrow some. "We feel like it's our responsibility to push back on that stuff," Zuckerberg said, "so oftentimes someone will come with a subpoena, and we'll go to court and say, 'We don't think this is enough.' Ultimately I think this stuff gets used for good."

Conversely, some governments fear Facebook's great database and the ease with which Facebook can be used to form networks and spread information. China has blocked the site since 2009. Iran, Pakistan and Saudi Arabia have all banned it at one point or another. Zuckerberg' girlfriend has family in China and they will be visiting and China represents almost a fifth of the world's population he's not reaching. But even without China, there's a distinct feeling of manifest destiny about Facebook. Plot its current growth on a curve and it hits a billion members in 2012. Is there a point at which all of them are on Facebook? Facebook's staff think it is totally possible and not just for the Internet but for all electronics we use becoming a conduit between us and people around us (GPS, telephones, iTunes, Ipods, etc.). For Zuckerberg Facebook penetration hinges on the ultimate extent of Internet penetration in the world, which in turn hinges on the adoption of smart phones in areas where Internet-connected computers are scarce. Facebook may enable people to stay in touch with others that you otherwise would not and the connections do not have to be satisfying or deep.

All technologies come with trade-offs, but for now Zuckerberg just doesn't seem that interested in the other side of the trade, the downside. There are some eloquent, persuasive critiques of life on Facebook out there, including Jaron Lanier's *You Are Not a Gadget* and MIT psychologist Sherry Turkle's forthcoming *Alone Together*. But they don't fuss him, particularly. "They're just looking at it through a completely different lens," he said. "And I appreciate that. Because it would be impossible for me to dissociate myself to that extent, to get that perspective. I mean, people write all kinds of different things, from 'It's the greatest thing that's ever existed' to 'It's the worst thing that's ever existed.'"

What looks like a meteoric rise to most people, he sees as an opening act. Because now that Facebook has scaled up to a species-level event, the real work can start: taking a

550 million – person network out on the highway and seeing what it can do. Zuckerberg could take the company public, but neither he nor the company needs the cash. It has never been about the money. "I think the next five years are going to be about building out this social platform," Zuckerberg said. "It's about the idea that most applications are going to become social, and most industries are going to be rethought in a way where social design and doing things with your friends is at the core of how these things work. If the last five years was the ramping up, I think that the next five years are going to be characterized by widespread acknowledgment by other industries that this is the way that stuff should be and will be better."

This won't make life any easier for people who aren't on Facebook. The bigger social networks get, the more pressure there is on everybody else to join them, which means that they tend to pick up speed as they grow, and to grow until they saturate their markets. It's going to get harder and harder to say no to Facebook and to the authentically wonderful things it brings, and the authentically awful things too. But while this happens, Zuckerberg is going to be growing too. The Zuckerberg who built Facebook won't be the same person as the Zuckerberg who runs it. He'll be getting older, traveling, maybe getting married, having kids, and as his life outside Facebook gets more complicated, maybe Facebook, the world he built in his own image, will get more complicated too: more sensitive to the richness that exists outside it, in the real world, and to the richness that passes through it in such enormous volumes every second of every day. For all its flaws, there was no other way for Facebook to begin.

Only someone like Zuckerberg, someone as brilliant and blinkered and self-confident and single-minded and social as he is, could have built it. "The craziest thing to me in all this," he said, "is that I remember having these conversations with my friends when I was in college. We would just sort of take it as an assumption that the world would get to the state where it is now. But, we figured, we're just college kids. Why were we the people who were most qualified to do that? I mean, that's crazy!"[10]

REFERENCES PROLIFERATION

1. *The Economist* (February 25, 2010). "Data, Data Everywhere."
2. *The Economist* (February 25, 2010). "Handling the Cornucopia."
3. *TIME Magazine* (Jan. 02, 1961). "Man of the Year: Men of the Year: U.S. Scientists."
4. *TIME Magazine* (January 3, 1969). "Nation: The Voyage: Poetry and Perfection."
5. *TIME Magazine* (January 3, 1983). "The Computer Moves In."
6. Isaacson, Walter (December 29, 1997). "Andrew Grove: Man of the YEAR" *TIME Magazine.*
7. Ramo, Joshua Cooper (December, 27, 1999). "Jeffrey Preston Bezos: 1999 Person of the Year" *TIME Magazine.*
8. Gibbs, Nancy (December 19, 2005). "The Good Samaritans" *TIME Magazine.*
9. Kher, Unmesh M. (December 25, 2006). "Person of the Year 2006 . . . In 2006 the World Wide Web became a tool

for bringing together the small contributions of millions of people and making them matter" *TIME Magazine.*

10. Grossman, Lev (December 15, 2010). "Mark Zuckerberg . . . Person of the Year" *Time Magazine.*

ECONOMICS AND BUSINESS

Economics for the purpose of this chapter will explore the technological impact on a national, regional, or global economy; i.e., macro economics or the branch of economics dealing with performance, structure, behavior, and decision-making of the entire economy. Macroeconomists study aggregated indicators (GDP, unemployment rates, and price indices) to understand how the entire economy functions. Models are developed to explain relationships between factors (national income, output, consumption, unemployment, inflation, savings, investment, international trade, and international finance). In contrast microeconomics focuses on actions of individual agents such as corporations and consumers and how their behavior determines prices and quantities in specific markets.

One could argue that capitalism was the winner of "The Cold War" since the Soviet Union dissolved in 1991. However the capitalism that has evolved in the former Soviet Republics and other countries throughout the world is quite different from what economist Adam Smith wrote "The Theory of Moral Sentiments" in 1759 when he coined the term 'the invisible hand' to describe the self-regulating nature of the free market. China for example has abandoned central economic planning for a type of state capitalism closely linked with nationalism with the 'old guard' Beijing communists still firmly in control. In the dawn of the 21st century some countries are moving toward market reform and others in the opposite direction. Europe is a combination of social democracy with a neoliberal economic system. The United States is an admixture of protectionism (socialism) and capitalism with politics as its underpinning. There is no convergence toward a single economic model. Social scientists generally agree that industrialization which produces the quality of life experienced in rich countries can be reproduced everywhere. The global economy is dramatically changing business models in a shrinking world and forcing societies onto a similar path of development.

According to "Globalization and Its Enemies" author Daniel Cohen thinks our ongoing wave of globalization is the third in a series beginning in the sixteenth century with the conquistadors, continuing in the nineteenth with the British imperial policies, and today occurring mostly in the realm of virtual reality leaving much of everyday life untouched. Unlike large scale movements of people to new lands in the nineteenth century the present phase involves mainly commodities and images and is immobile. Immigrants in 1913 made up 10% of the world's population and today make up 3%. Cohen argues most trade is between rich countries . . . as in Europe (which comprises about forty percent of global commerce) where about 66% of its imports and exports are within Europe itself. He posits that technology is unevenly diffused and the poor stay poor. The customers of most local African businessmen cannot afford price increases that will pay for the business to computerize. Poor countries stay poor because they have few products the rich countries want or need. It is not that they are exploited but that they are neglected and forgotten. The press and television are continually showing them images of the wealth that they lack. Today's globalization universalizes the demand for a better life without providing the means to obtain it.[1]

The global warming being experienced worldwide has a relationship to the extraction and consumption of hydrocarbons which is integral to industrialization. The observed rapid Antarctic ice cap melting and retreating ice flows in Greenland are clear indications that a climate shift is underway that could severely effect the way we live. A global problem requires a global solution that requires global cooperation among the world's nation-states. Other then governmental pronouncements and limited voluntary efforts there has been no serious agreement to reduce hydrocarbon release into the atmosphere. If coastal cities and large areas of arable land are flooded . . . the movement of millions of people plus dwindling resources of food, water, and energy may cause humanity to act in a global unified way or push humanity to an even more insular posture of only taking care of its own. Global warning is a byproduct of globalization and despite moves in some countries to develop an admixture of energy supplies the world is still overly dependent on dwindling reserves of oil and natural gas that continue to spew hydrocarbons into the atmosphere. The two most populous countries on Earth, China and India are transforming rapidly into industrialized behemoths thus increasing the demand for oil and gas at an alarming rate. Instead of reducing scarcity worldwide industrialization is reproducing the resource conflicts of the past on a much larger scale. Extending the energy-intensive lifestyle of the rich world to the rest of the world is not only increasing scarcity to the have-nots, it is destabilizing the environment. Technology can come to the rescue of humanity by developing less dangerous sources of energy. We cannot stop climate change, we can only slow it down. Cohen distinguishes two kinds of economic growth: 1. Smithian reflecting Adam Smith's version in "The Wealth of Nations" where growth is obtained by utilizing the benefits of labor; 2. Schumpeterian reflecting being driven by continuous technological innovation. The Smithian variety can exhaust itself while Schumpeterian growth according to Cohen is *apriori* without limit.[1]

Suzanne Berger is skeptical in "How We Compete" regarding the claims that different models of economic development ultimately converge. She writes that there is broad agreement on the fundamental forces driving globalization: freeing of trade and capital inflows; deregulation; decreasing cost of communication and transportation; low wage workers and engineers in various countries; a revolution in information technology making it possible to digitize the boundaries between design, manufacturing, marketing, and to locate these functions in different places. Berger posits that however securely established it may seem, globalization is not irreversible and over time its disruptive effects tend to result in de-globalization. Using a five-year study by MIT Industrial Performance Center, Berger provides ample evidence that five hundred international companies survive and prosper in the global market by utilizing different strategies. For example Samsung makes almost everything itself while Dell strongly focuses on distribution and outsourcing all manufacturing components. Faced with similar challenges companies thrive or fail in different ways. Berger does not totally reject the view that as old jobs are lost new technologies and industries will appear to replace them but points to the trend that laid off workers many times accept large reductions in pay in order to work. Employing cheap labor abroad by outsourcing poses a real risk to employees. Companies whose main asset is cheap labor cannot match companies that succeed over time through long-term working relationships with customers and suppliers plus their specialized skill sets. Capitalism comes in several varieties reflecting different cultural traditions and political systems: 1. "liberal market economies" such as in Britain and the United States in which allocation and coordination of resources takes place mainly through markets; 2. "coordinated market economies" like Germany and Japan where negotiation, long-term relationships, and other nonmarket mechanisms are used to resolve major issues. These two different systems compete and learn from each other resulting in cross-fertilization, not evolution toward a 'one size fits all' single model. What works varies amongst both companies and countries. It is historically false to believe that there is only one way to succeed in business. Berger summarizes the results of years of research: "Succeeding in a world of global competition is a matter of choices, not a matter of searching for the one best way – we discovered no misconception about globalization more dangerous than this illusion of certainty."[2]

Cell Phones, Electric, Water

In underdeveloped countries it is not uncommon to walk many miles only to find the person or object desired is not available. A New York investment banker Iqbal Quadir seized on the idea that "a telephone is a weapon against poverty." Knowing very little about the telecom industry he was determined to make telephones widely available in his Bangladesh homeland. Inspired by the Bangladeshi Grameen Bank known for supplying 'microcredit' or small loans he formed a consortium with Grameen Bank and Telenor, a Norwegian mobile phone operator that provided the required telecoms expertise. GrameenPhone began services in 1997 and by 2006 had more than six million subscribers of the more than nine million subscribers throughout the country with six mobile operators. To accommodate the total population of about eighty million people 'telephone ladies' provide 'village phones' in the country's fifty thousand villages by sharing their phones. Although a small portion of the total mobile phones in circulation the 'village phones' account for about 30% of the network traffic due to the sharing amongst a large number of users. Bangladesh's GDP has increased more from the proliferation of mobile phones than repeated infusions of foreign aid because they promote economic activity, prevent wasted journeys, make it easier to look for work, and widen access to markets. This approach is being replicated in Asian and African countries. Mr. Quadir argues in favor of using technology to solve chronic problems of poverty and economic development. Traditional governmental aid increases the gap between politicians and the people. Putting the power in the hands and brains of the people is the key to economic success.

There are historical precedents for this bottom-up approach to economic development. The inventions of spectacles, water wheels, clocks, and printing in medieval Europe empowered the people from below thus stimulating economic development even though the church and state opposed it. The Industrial Revolution was the result of bottom-up activity, not government planning. Mr. Quadir is looking to apply his telephonic success to electricity by establishing small, local power plants to provide electricity to a handful of homes, shops, and businesses; he has teamed up with Dean Kamen, the American inventor of the Segway electric scooter. The generators can be powered by biogas extracted from cow manure. One entrepreneur is funded with a microcredit loan and turns manure into methane gas and fertilizer. Another entrepreneur funded with microcredit buys the methane to power the generator, and sells the resulting electricity. Each generator is able to power up to twenty households or shops allowing businesses to work and stay open longer, students to study longer, and people to enjoy an old electronic device called television and other forms of entertainment. The next step is to mass produce the generators, hopefully in Bangladesh to further enhance the economy and provide even more jobs to the production workers and the repair workers.

Mr Quadir is pursuing two other bottom-up initiatives. CleanWater, is dedicated to supplying safe drinking water to Bangladeshi villages, where arsenic contamination is a deathly problem. He hopes to license a chemical preparation that can remove arsenic from water and make it safe to drink. The chemical would then be distributed and sold, like salt, via a network of local entrepreneurs; it is estimated that the cost would be well within the reach of most villagers (about $3.00 per person per year) to ensure a safe water supply. The second initiative is called CellBazaar. The idea is to create an electronic marketplace that can be accessed via mobile phones . . . a phone-based equivalent of newspaper classified advertisements. If someone wants to sell their electric fan, for example, they can list it in CellBazaar. It will make price information more transparent and widely available. Keeping the system simple is a goal so it will not handle transactions but will simply put buyers and sellers together via their mobile phones. Contact can also be made by texting. Electronic commerce has considerable appeal in countries where the transport infrastructure is often inadequate.

Mr Quadir believes that just like economists invoke the 'invisible hand' of the market he relates to technology as the 'invisible leg' of the market that can move an economy from one state to another. "Without necessarily introducing enlightenment or new arguments, technology can quietly initiate novel ways of making things or trading them, potentially redistributing economic and political clout." Even when government adopts a sensible policy there is no guarantee of it being implemented. The Bangladesh government had a policy promoting universal access to telecommunications but nothing really happened until the establishment of GrameenPhone which prompted the government to issue more mobile licenses thus leading to a thriving market. Technology makes it possible in Bangladesh's experience to change the facts on the ground first, so that government policy can then follow.[3]

Internet Searching

A huge worldwide multi-billion dollar Internet business is called Search Engine Optimization (SEO). Search engines are constantly testing and implementing new algorithms to optimize their searches because achieving a high search ranking is crucial to success for many online businesses, hungry for website visibility and website visitation. Google, the largest Internet search engine is constantly sending out its spider 'bots' to crawl around the Internet to scrutinize billions of web pages and determine their Google rankings in search results. This "Google Dance" takes place two or three times every year and are given names like Bourbon, Gilligan, and Jagger. The Google Dance can scare companies since rapid and often temporary falls down the pecking order of search placement are common. SEO experts analyze algorithm updates to test an array of potentials to see what works and what does not work. By applying what works to clients' websites to enhance its positive characteristics that search engines consider positive, the SEO expert can boost their clients' online rankings. The expert tinkers with the website using various techniques such as simplifying complicated page addresses, rewriting copy to produce single-theme pages with more accurate titles, adding keywords to the invisible page descriptions (or 'megatags') read by indexing software, and putting product information stored in databases directly onto fixed pages, so that search engines' bots can read it. In the past new algorithms could be ascertained by combing through online patent applications. Google like its competitors does its best to control access to its secrets. One method is to no longer file for patents because they eventually become public. Firms are moving from monthly pay-per-click advertising to SEO companies to improve their rankings in Internet searches.

A site's popularity is gauged by the number of incoming referral links . . . considered to be the most powerful determinant of a web page's importance. Links from directory sites, link swapping with prominent sites, and blogger links are solicited by site optimisers (i.e., SEO experts or SEO companies). Unethical methods, known as 'black-hat SEO,' may rent links on long-established sites or even exploit loopholes in website-management tools to place hidden links on prestigious sites such as those maintained by universities. The 'link farm' technique spams the web with automatically generated bogus blogs (called 'splogs') that link to a website to give it a boost. Copied or 'scraped' content from legitimate sites is often used in link-bearing splogs to fool the spider bots. Indexing bots may be fooled by 'cloaking' which presents content that is different from what surfers to the site will actually see. 'Keyword stuffing' hides popular search terms on a page to attract visitors (e.g., using white text on a white background). Algorithm updates have improved their ability to keep cheaters out. The financial stakes are high and SEO firms will continue to seek new ways to boost their clients' rankings.[4]

Internet listening

One of the Internet's appealing features is being able to locate, read, listen, and participate on just about any topic or subject, no matter how common or obscure. There seems to always be at least one website, blog, or discussion forum where people come together in cyberspace to discuss the topic. There is a site for just about every topic from the best software platform to the most obscure forms of Japanese poetry. Internet companies have archived discussions that go back to the earliest days of the Internet in the 1970's (Usenet reached via Google dates back to 1979). Online debates have become a very popular forum by appending web-based discussion boards, and blog comments. These online discussions have very specific topics and generally have little interest to the majority of Internet users but the unedited, brutally honest nature of these discussions is critical to large companies looking for trends, and finding out what customers really think . . . especially about them. Many firms monitor online chatter as an adjunct to more traditional forms of market research.

The American food giant ConAgra (butterball turkeys, Healthy Choice meals, etc.), tracks discussion groups to keep abreast of new diet trends; e.g., Atkins and organic food. It is viewed by marketers as an inexpensive fast way to determine if something is a real trend or just a fad. From information gathered from these discussion groups conventional market-research techniques can then be utilized to test an 'overheard' hypothesis. Public-relations teams and some company executives listen-in to online discussions to stay in touch with the public mood and note any trends that can be passed on to the company's marketers. Listening in the 'old fashion' way, by individually going on line to a particular site is being replaced by computing power to make it more systemic. Accenture and IBM have systems that trawl the web in search of trends and insights. These programs need to be periodically updated so as to interpret all the jargon, abrreviations, and slang that is commonly used. Smaller specialty firms have sprung up using a combination of computers and human researchers to track discussions and spot trends early. They identify members of online communities who are most likely to influence other participants. Some of the world's largest companies use these services (e.g., General Motors, Ford, and Microsoft).

Non-Internet users, of course are excluded from this type of research so telephone surveys and focus groups continue to be useful for marketing. The significant advantages of 'listening on the Internet for profit' is the high speed and low cost. Traditional market research can take weeks or months before the results are tabulated and finalized. Participants in discussion groups can say whatever they want, thereby providing broader information whereas people answering survey questions only respond to what the surveyor thinks to ask. Sometimes the results are surprising when people's perceptions are opposite to what the company assumes. Privacy like so much on the Internet is a big concern. Many companies are reluctant to discuss their activities around 'Internet listening.' Participant consent is not given when tracking public forums for profit and participants often do not know their conversations are being listened to and analyzed. Posting product plugs on forums is generally considered to be unacceptable but it certainly can be done. Monitoring of discussion groups could provide an answer to consumers who complain that companies do not listen to them.[5]

Video Advertising

The Internet, hundreds of television channels, and the ubiquitous video camera make it all the more difficult to reach the consuming public. Advertisers are doing their best to go where they have the best chance of reaching people: supermarkets, gas stations, retail stores. 'Digital signage' is the new name for flat-panel displays showing a rotating series of advertisements mixed with news and entertainment. Digital signage is placed in 'high traffic' (i.e., busy with people) areas of malls, shops, and gas stations. The signage can be updated by satellite or Internet links. Updating allows to vary the advertisements shown depending on the time of day, season, or local factors such as demographics or weather. The ability to reach consumers just as they are deliberating about which item to pick from the shelves has potential. Point of Purchase Advertising International, an independent trade association, estimates that more than 70% of purchasing decisions are made in shops; for example most people write 'toothpaste' or 'shampoo' on their shopping list rather than a particular brand. As the prices on Internet links and large flat screens drop, retailers and suppliers around the world are moving to this form of advertising.

According to Premier Retail Networks (PRN), a firm that operates Wal-Mart TV to more than 2,500 American stores, the network has over 50 million viewers a month exceeding the viewership of the most popular television shows. The network carries a sophisticated multi-channel offering: screens positioned in the best points around the store show advertising picked to suit individual departments, while other screens provide entertainment . . . interspersed with even more advertising . . . to customers standing in line to cash out. The TV displays customize their content often from standard television advertisements; this is a common practice with digital-signage installations. It makes economic sense to recycle their costly television advertisements for a captive audience in the process of buying things. Widespread utilization of digital signage is growing slowly since the cost of installing and running (creating and managing fresh and effective content) the network is not cheap especially for the large chain stores. Its ability to reach customers as they shop gives it a long term edge over the traditional forms of advertising, such as television, radio, and billboards.[6]

Television Place Shifting

Television 'timeshifting' began in the 1970's with the advent of the video recorder that enabled people to watch programs whenever they wanted to; a form of consumer empowerment since the broadcasters could no longer control the time that everyone would watch their broadcasts. Timeshifting really came into its own with the advent of personal video recorders (PVRs) such as the TiVo. Such boxes can be easily set up to record your favorite programs whenever they come on, and their large storage capacity allows dozens of shows to be stored for later playback. Advertisers do not approve of this technology since it allows people to skip over their advertisements.

Just as timeshifting allows viewers to choose when to watch something, 'placeshifting' lets them decide where. For a long time people have been able to carry recorded shows (on videotapes or DVDs) around with them. Now the placeshifting of live broadcasts, as well as recorded shows, is available. We can have any content, anywhere, anytime, on whatever device we own. This placeshifting is part of a broad personalization trend for media consumption: Music playlists, mobile phone ringtones, television programming, and movies.

Placeshifting in television started with a device placed on top of a person's television to gather video feeds from cable or satellite set-top boxes, DVD players and PVR's. By re-transmitting a video stream via a broadband Internet connection the user can view whatever he/she desires from anywhere in the world. The genre was introduced in

2003 with the TV2ME device. Other devices include: Sling Media's Slingbox, Sony's LocationFree, and Monsoon HAVA. Several companies have also developed computer software that allows consumers to placeshift media stored on their personal computers (PC) to a remote device. Placeshifting has a narrower appeal than the impact of timeshifting. Travelers, late night workers, and weekend workers who want to watch their home sport teams or favorite programs do not constitute a mass market presently. Placeshifting offers benefits to the Television and film industries by vastly increasing the number of viewing screens by turning any PC or handset with a broadband connection into a potential television. However, the downside for business is that it makes it more difficult to charge for new streaming video and mobile-TV services. Why pay extra for a limited selection of programs when you can stream TV from home to your computer or handset. The technology circumvents national censorship and regulatory rules and flies in the face of national rights to television shows and sporting events, or regional release windows for films. If nothing else placeshifting is hastening the arrival of "when I want, where I want" television.[7]

Business Information Revolution

National Cash Register (NCR) revolutionized business when in 1879 James Ritty received a patent for a wooden contraption that he dubbed the "incorruptible cashier." NCR sold the device, which was little more than a mechanical adding machine to businesses becoming the first widely used mechanism to manage information flows in America. Through an automatic bell ringing every time the cash drawer opened it reduced theft and recorded every transaction providing an instant overview of sales. Sales data remains a very important asset to any business. With enormous databases huge corporations like Wal-Mart are able to run programs that show trends such as purchases in bad weather, in nice weather, or any time . . . day or night . . . seven days a week. Ironically NCR has used its data-warehouse sharing unit to show large customers such as Wal-Mart what is popular purchases and when they are popular purchases. Unexpected outcomes do happen and allow the conglomerate to insure certain items are always available at particular times based on trend analysis. As the price of these 'business technologies' continue to drop with the price of computing, storage, and software, they have become much more mainstream. Companies are linking their systems and collecting more data than ever before. Using 'data-mining' techniques they are able to get a complete picture of their operations in a very understandable format. By so doing the firms are able to operate more efficiently, see trends, and improve their forecasts. An excellent example of successful 'data mining' is the Swiss telecom operator Cablecom that reduced customer defections each year from twenty percent to under five percent. Its software showed that nine months into their subscription customers were making their decisions to leave as shown by things such as the number of calls to customer support services with customer defections peaking in the thirteenth month. Seven months into their subscription select customers were offered special deals and the defection rates dropped dramatically. Best Buy a chain retailer found that seven percent of its customers accounted for forty-three percent of its sales . . . it reorganized its stores to concentrate on the select seven percent. Analytic techniques uncovered the best predictor that a passenger would catch his/her booked flight: he/she would have ordered a vegetarian meal.

Analytics is rapidly becoming big business because it can forecast or uncover correlations that have increased profits. By analyzing past sales data opera and theater companies can increase attendance through a better informed marketing campaign (the British Royal Shakespeare Company analyzed seven years of sales data for a marketing campaign that increased regular visitors by 70%). The first step in analytics is to improve the accuracy of the company's information which is not always easy to do. When a company has not been minding its information collection warehouse it often has obsolete, duplicated, inaccurate, or incomplete records. A common error is that names might be abbreviated in one record but spelled out in another, leading to double-counting. This experience is common to those of us with large address books in our Internet browsers where sometimes names are listed two or three times as Jim Smith, Jimmy Smith, and James Smith. As sensors are increasing rapidly in everything from traffic flows to patient blood flows in hospitals many businesses see business intelligence as a key to the future of the company (e.g., IBM has invested $12 billion since 2006 to open six analytics centers with four thousand employees worldwide).

Many corporations and companies have database problems . . . mostly because of poor quality and many managers do not trust information that is analyzed from its database. Managers do not want more hay from their database haystacks, they want to find the golden needles. It is projected that as analytical techniques grow and become more reliable that decision making will increase based on computer generated algorithms. Individual hunches that oppose the computer analytics will eventually not be taken seriously . . . this will occur when the reliability of the analytics is no longer questioned.

Stored information about past transactions is called 'dead data.' More and more companies are moving from past informational correlations to analyzing real-time information flows. Wal-Mart whose revenues are greater than the GDP of many countries has thousands of stores worldwide, several million employees, and hundreds of millions of transactions every week. To manage such enormous information, let alone to make sense of it can be quite daunting. However its inventory-management system called Retail Link enbles suppliers to know the exact number of their products on every shelf in every store at any given moment of time plus the rate of their products sales (by hour, by day, over the past year, and more). It not only gives information on when and how their product is selling but also on what other products in the shopping cart it is selling with. Through this software Wal-Mart was able to change its retail business model by leaving most of its stock management in the hands of its suppliers allowing it to shed inventory risk and reduce costs. One of the world's largest supply-chain operators Li & Fung does not own any factories or equipment. Through 'real-time' information it directs a network of thousands of suppliers in forty countries. Li & Fung dealt with clients in the past by phone and fax with e-mail considered as its high technology. With a new web-services platform it speedily processes bids solicited from pre-qualified suppliers. Agents audit factories in real time on hand-held computers. Clients can monitor every stage of an order from initial production to shipping. Real time information flows have greatly increased the amount of data being collected. Such vast information is able to forecast when machines will break down enabling firms to act before it happens. The information system allows companies like Li & Fung to identify trends: it moved its southern China production north when a shortage of workers and new legislation raised labor costs. It also received advance warning of the economic crisis, and later the recovery, from retailers' orders before the trends became apparent. Investment analysts use country information provided by Li & Fung to gain insights into macroeconomic patterns.

The enormous amounts of data that can now be shared, analyzed, and stored has allowed videoconferencing to become another valuable business tool. It allows buyers and manufacturers to examine the details of a product (color, stitching, quality, size, etc.). Real-time images allow changes to be made faster and more efficiently. 'Predictive analytics' can transform health care. Research is being conducted to spot potentially fatal infections in premature babies by monitoring subtle changes in seven streams of real-time data, such as respiration, heart rate and blood pressure. The monitoring of the heart rate through the electrocardiogram generates 1,000 readings per second. In the past this kind of medical information was printed out and examined periodically. By digitizing and analyzing the information in real time medical personnel can detect the onset of an infection even before obvious symptoms appear. Our human eyes cannot see it but our machines can.

Two technology trends fueling the new use of data is cloud computing and open-source software. Cloud computing uses the Internet as a platform to collect, store and process data; it allows businesses to lease computing power when they need it eliminating the need to purchase expensive hardware. Amazon, Google, and Microsoft are leading the way in making their computing infrastructure available to paying customers. By seeing patterns across the whole of the business (e.g., human resources and sales) the information can easily be shared. Special programming languages and software allows ordinary PC's to analyze what in the recent past was only able to be accomplished with supercomputers. The New York Times used cloud computing to convert 400,000 scanned images from its archives (1851 to 1922) in thirty-six hours. Visa, a credit card company crunched seventy-three billion transactions (thirty-six terabytes of data) in thirteen minutes when traditional methods would have required thirty days.[8]

Clicking for Gold

Companies are compiling massive amounts of data across the Internet economy about people including activities, likes and dislikes, relationships, and locations at any given moment. The big Internet companies like Facebook, Google, eBay, and Amazon do not reveal very much about their massive databases on people and how they are using them. The reticence partly reflects fears about consumer unease and unwelcome attention from regulators. Users, regulators, and elected officials are becoming more anxious about how the information is being utilized. Disclosure could impress and/or alarm users regarding their privacy. Part of the silence probably comes from the fear of losing their competitive advantage, especially if they have developed software only imagined in science fiction. In a expoentially growing information age, data is key to profitability and those companies that understand this have a big lead over those who do not. Their silence is not about privacy but about revealing valuable trade secrets to competitors. From a profitablity standpoint it makes sense that a firm like Amazon would want to know what books you purchase, what books you browse, and how long you spend reading each page of your electronic books on Kindle or other devices.

Traditional business collects customer information from purchases and surveys. Internet companies gather information from everything that happens on their sites. Using this information has been going on longer than the dawn of the 21st century and has become an Internet company's biggest marketing treasure. Large firms run controlled experiments or pilot tests to see what works best. They use a statistical technique called 'collaborative filtering' to make recommendations to users based on what other users like. For example about sixty-six percent of Netflix's customer film selections come from computer referrals. Ebay is constantly making adjustments based on listing activity, bidding behavior, pricing trends, search terms, and the length of time users spend on a page. Every product category is actively managed. Google is the leader in mining large amounts of information for economic value. It utilizes data exhaust (by-product of user interactions) to exploit information and recycle it to improve the service or create new products. Google's innovation thanks to one of its founders, Larry Page changed the way Internet searches are conducted. Instead of counting the number of times a word appeared on a webpage to determine its relevance (wide open to manipulation), the number of inbound links from other webpages was counted. Like the value of a book increases proportionately to the number of times it is cited by other souces so the more links suggest a webpage is more useful. Hackers quickly learned how to abuse Google's innovation with 'link spam' so Google analyzed the data and came up with the solution. Users often only want one page of the thousands or millions a search might reveal. By their selection of that one page the algorithm is reset to feed it back into the service automatically. Ergo Google is a huge data-mining business. This technique is not new but in the past was slow and laborious since data had to be compared manually to find the best and most commonly used items/issues.

Google gave new meaning to recursive learning when it applied the principle from its various services to give the world the best spell-checker in almost every language. The program developed is based on all the misspellings users type into a search window and then correct by clicking on the right result that appears. With billions of queries daily the results grow quickly. Other search engines in the 1990s had the opportunity to do the same but did not pursue it. By mining the data exhaust Google collected the gold dust in the dross. Google is using the same approach with two of the most difficult computer venues: translation and voice recognition. Both have been big stumbling blocks for computer scientists working on artificial intelligence. In the past scientists were attempting to program all of a language's rules and exceptions. Google saw it as a big math problem that could be solved with a lot of data and processing power.

For translation, it took from its other services. Its search system had copies of European Commission documents, which are translated into around 20 languages. Its book-scanning project had thousands of titles that had been translated into many languages. The translations are of the highest quality, completed by experts to precise standards. Rather than teaching its computers the rules of a language, Google provided the texts to make statistical inferences. Google Translate is growing covering more than fifty languages in 2010. The system identifies which word or phrase in one language is the most likely equivalent in a second language. If direct translations are not available (e.g., Japanese to Hindi), then English is used as a bridge. IBM tried this method unsuccessfully in the 1990's with a French-English program because its database was not large enough. The design of the feedback loop is critical. A translation start-up in Germany called Linguee is trying something different: it presents users with snippets of possible translations and asks them to click on the best. That provides feedback on which version is the most accurate.

Data exhaust is critically important with voice recognition. To use Google's telephone directory or audio car navigation service, users dial a number and say what they want. The system repeats what it hears and when the user confirms it, or repeats the query, the system develops a record of the different ways the target word can be spoken. It computes probabilities and does not learn to understand voice. Google keeps the data from voice queries allowing it to constantly improve its voice-recognition system

The re-use of data is the new model for how computing is done. Huge data sets allow this to happen and advance rapidly. Statistical analysis is the way of the machines, not understanding. Many Internet companies are now understanding that 'understanding' is not the future. Facebook regularly examines its huge databases to boost usage. Zynga, an online games company, tracks its 100 million unique players each month to improve its games. The successful companies have built a culture to deal with enormous amounts of data that traditional companies just do not have. Recycling data exhaust is a common theme at Google and most of their projects are being tested; they are trully in continuous development. Google users can store their medical records and it might allow the company to spot valuable patterns about diseases and treatments. Another service allows users to monitor their use of electricity, device by device. It could become an enormous database of household appliances and consumer electronics possibly forecasting breakdowns. The aggregated search queries, which are free on the Internet are used as accurate predictors for everything from retail sales to flu outbreaks. Google's mission is "organize the world's information." Google claims it does not need to own the data, it desires to have access to it and keep it away from competitors. Google is working on a new initiative that will reset all its services so that users can discontinue them easily and take their data. By reducing the 'barriers to exit' in an industry built on locking in the customer, Google would become once again the innovator that leads its industry. Will users be more inclined to share their information with Google if they know that they can easily take it back?[9]

U.S. Stock Market Plunge

On May 6, 2010 the United States stock market spun out of control between 2:40 and 3:00 PM and took a record-setting ride plunging down (the Dow Jones average dropped 1,000 points in minutes) only to abruptly snap partway back. The BATS Exchange, a large electronic exchange based near Kansas City prevented more than 47 million orders (95% of all orders) from executing during the panic. The BATS exchange rejects orders if the price would be more than five percent or $.50 away from the last completed transaction. This brief frenzy of electronic trading has experts and market operators very concerned. There is a surprising consensus about what should be done but nothing has. The market is increasingly fragmented and computerized and federal regulators have not shown any urgency in rewriting the rules that governs the market. Regulators, traders, and academics all agree that the gist of the solution is that markets need uniform rules for intervening when a stock goes into free fall.

The connotation that the 'stock market' is a single entity is a misnomer since investors can buy and sell stocks through about fifty markets in the United States. Most trades are placed via computer networks at the direction of computer programs. Orders are routed automatically to the market offering the best price. If computerized sellers cannot find enough computerized buyers the system can spin out of control. Dendreon, a Seattle biotechnology company, saw its stock price plunge sixty-nine percent in seventy seconds on April 28, 2009 before trading was halted. Most of the loss was instantly erased when trading resumed the next day. Because these declines can reflect a temporary shortage of buyers rather than a permanent loss of value, some of the markets impose 'circuit breakers' to stop trading and protect sellers from taking unnecessary losses. On May 6 the New York Stock Exchange briefly suspended trading in some shares then slowed the pace of trading to give sellers a better chance to find buyers.

Market safeguards make sense only if they are applied uniformly . . . on May 6th sellers moved to other exchanges with fewer restrictions. Some of the downward spirals ended selling at one penny because in some cases, the supply of buyers on those exchanges already had been exhausted, causing the computerized trading programs to offer shares at lower and lower prices. Part of the uniform solution being called for is to change any one exchange having an independent circuit breaker. The Securities and Exchange Company that oversees the nation's equity markets requires a suspension of trading in the event of a broad market collapse (defined as a drop of at least ten percent in the Dow Jones industrial average, which is based on the share prices of thirty large American companies). Many believe that computerized trading is not properly regulated and could be improved simply by requiring all sellers to specify a minimum price below which they do not want to complete the sale of their shares. Market orders, placed at the best available price, can be too risky in the fast-moving age of electronic trading. The May 6th panic could have been avoided if sellers had placed 'limit orders' (i.e., fixed minimum price below which there is no sale) instead of 'market orders.'

Smart Systems and the Economy

The convergence of the physical and digital worlds has spawned companies interested in making profit from smart systems and those that succeed will disrupt more than one industry and probably the entire economy. Smart systems allow for a plethora of new services and business models. Utilities are not the only sector that is and will continue to benefit. Linking office equipment with the computer can control everything from air conditioners to automatic vacuum cleaners and air fresheners. One company in the paper industry achieved a five percent production increase by automatically adjusting the shape and intensity of the flames that heat the kilns for the lime used to coat paper. One company in the food industry allows food suppliers to tag and trace their wares all along the food chain and consumers to check where they came from. The health and location of livestock is being monitored by implanted sensors in ears. Maintaining roads and equipment will become much more efficient thanks to detailed digital maps. Highways can be reviewed digitally including what is underground. By knowing the location and maintenance history of equipment repair can be conducted prior to any breakdown. In the transition to the 'new world of smart systems' the 'old world' and its equipment will be made to work more efficiently. A company based in Dubai may be the beginning to the 'new world of smart systems.' Pacific Control operates a global command center and remotely monitors buildings, airports, and hotels for energy use, security, and equipment.

Once devices are connected their use can be metered. Metering equipment has spawned an industry renting equipment rather than making expensive sales. Rolls-Royce which makes expensive aircraft engines, rents them out to airlines, billing them for the time that they run and telling them when they may fail and should be serviced or replaced. Makers of blood-testing equipment have taken to charging only if the device actually produces usable data. And Joy Mining Machinery, a maker of mining equipment, charges for support by the ton. Some firms are using metering in innovative ways. Zipcar and other car-sharing firms put wireless devices with sensors into their vehicles so that customers can hire them by the hour. Insurance firms (Progressive and Coverbox) request the installation of car smart systems to measure for how long, how fast, and even where a car is driven thus basing insurance premiums based on individual drivers' behavior rather than age and sex. Machines are having a much closer relationship to customers. Japanese vending machines can recognize a customer's age and sex and change its displayed message accordingly.

Hewlett-Packard has a goal to increase demand for its hardware by scattering millions of sensors around the world and at the same time offer additional services based on networks of sensors. The success of the iPhone is transforming Apple into a service and data business. Apple did not anticipate that the applications that run on its smartphones ('apps') would become so incredibly profitable. Apps are being downloaded by the billions. Apple is making money from all the data it collects. iAd, a mobile platform for advertising is continuing this profitability. Some firms will make a living based entirely on mining 'data exhaust,' the bits and bytes produced by other activities. Google's PowerMeter lets users check online use of electricity and gives the company access to more data to analyze and sell advertisements against. Sensors can monitor almost everything and discussions are taking place to tax things like pollution and changing behavior by manipulating charges for public goods.

YouTube made video-sharing very popular because it converted all videos to a common format. A common format for sensor data feeds is critical in order to seriously change the world of smart systems. With a common format, alerts, data storage, and visualization tools will be accessible to all users. Wikitude and its smartphone app 'World Browser' may be the beginning of serious world change. World Browser checks the device's location and the direction the camera is pointing and then overlays it with notes from other users about the visuals like landmarks and buildings. In principle this could morph into a collaborative annotation of the entire physical world including other people. 'Augmented ID' is now available that uses facial recognition to display information about a person shown on a smartphone screen.

Three trends stand out regarding smart systems and the economy: 1. Providing better information should lead to improved pricing and resource allocation; 2. Integrating the virtual with the real world will rapidly increase the move from physical goods to services and more things will be hired and not purchased; 3. Data and algorithms to analyze data will become the new economic value. Data and its inherent knowledge may become a factor of production just like land, labor, and capital. Companies and governments will become very protective of their data assets.[10]

Diane Coyle in her seminal book "Weightless World" may have predicted our future . . . a future in which bytes are the only currency and the things that shape our lives have no weight. Economically, information and services have been around for centuries and are distinctly different from agriculture and manufacturing because what they produce is weightless; i.e., have material embodiments (even information must be physical) but their value is not related to their mass. The shift to weightlessness Coyle claims portends the end of secure long-term, even life-time employment. The distribution of income in the weightless world will be highly skewed toward the weightless sectors creating a socio-political crisis: 1. Existing institutions of the post-war welfare states will fail to serve the increasing numbers of people who are neither full-time bread winners or not working at all; 2. Political support for the inclusive welfare state continues to diminish; particularly for those who benefit from the higher, rising inequality; 3. Public social insurance becomes even more necessary. Coyle postulates that the highly productive future economies will benefit a small minority with the majority forced to work for stagnant incomes and lead unpredicatble, insecure lives. One of three outcomes can occur: 1. The majority will have the system imposed upon them; 2. The majority will impose something in their favor; 3. The majority will be given a stake in the system. In the United States with the largest prison system and private police forces in the world the majority is having the system imposed upon them. Only the third alternative (the compromise) offers any hope of a peaceful transition . . . the kind of systemic public good that historically the markets cannot provide. In the information age all customers are potential competitors since many kinds of information have 'positive network externalities' i.e., the benefit of owning a particular piece of information (e.g., an operating system) increases with the number of others who also own it. Intellectural property rights allow one party to capture the benefits (if any) from producing information. As government-created and enforced monopolies they suffer from all the issues that have always afflicted monopolies; i.e., the monopolists will do their best to legally screw the public. In short an information economy is almost guaranteed to experience market failures of one sort or another.[11]

For most IT (information technology) firms smart systems simply means more business and the sector will probably see the least change in the new economic information age. It is estimated that the decade of the teens (2010 to 2019) smart computing technology will represent about fifty percent of the spending on IT equipment and software in America. Internet-enabled devices stand to greatly benefit and wireless sensors are growing very rapidly; they are expected to continue growing by close to 1,000 percent by the middle of the decade. The data collected is also increasing rapidly so storage is also a huge growth industry. Business analytics software and businesses that sift through mountains of data are also likely to be hot commodities. Interaction with the physical world that we inhabit is critical and many new programs will be needed. New startups are everywhere offering to help businesses stay green and cut greenhouse gases. Platforms to integrate the data streams from all kinds of sensors are another new market. Palantiri Systems for example is considered by some to be a "Facebook for sensors." Their service allows devices to have their own 'page' on corporate social networks and people can ask questions about them as their readings are shown in a newsfeed form thus making the devices part of the discussion. IT services are expected to make the most money particularly setting up smart systems in cities (Big IT firms are already helping China with its enormous urban growth with the expectation that as more people migrate to its cities the demand will only increase accordingly).[12]

Economic Impact on Higher Education

The information age is also having a dramatic impact on higher education especially in England and America. British universities including Oxford and Cambridge have always been under a system of state control that is now making more demands based on economic principles. Private American firms such as IBM, Oracle, and SAP have sold the British government, its bureaucracies, including its universities intensive management systems that make use of information technology (IT). From the late 1980's the system fostered by both major political parties insists that research 'output' be delivered with a speed and reliability like in the corporate world. The research is expected to somehow be useful to both the public and private sectors thus enhancing performance in the global marketplace.[13] Over the last several decades the most obscure management practice, the 'Balanced Scorecard' (BSC) has had the greatest impact on British academic life. Published in the Harvard Business Review in 1992 by Robert Kaplan and David Norton the BSC focuses heavily on the setting up, targeting, and measurement of statistical 'Key Performance Indicators' (KPI's). With networked computer systems it is possible to expand the number and variety of KPI's. The Balanced Scorecard focuses on four business areas: 1. relations with customers; 2. internal business process; 3. financial indicators (profit and loss); 4. innovation and learning.[14]

The pharmaceutical industry is a major segment of the British economy that is an intensive user of university research. The British government's invitation to business 'end users' to take a prominent part in the evaluation of academic research is alarming given the pharmaceutical industry's abuse of research integrity in the interest of its own profit. GlaxoSmithKlein made secret payments to an academic psychiatrist to promote the company's drugs and suppressed unfavorable research on its top-selling drug

Paxil; it paid a fine that dwarfed its profits to settle charges of consumer fraud.[14]

The central government through the United Kingdom Treasury decides the outlines of policy . . . the amount of money to be distributed to universities for research and how research excellence is defined which determines the allocation. Administrative detail is handled by the special state bureaucracy the Higher Education Funding Council for England (HEFCE). The HEFCE control system 'Research Assessment Exercise' (RAE) does a review process every six or seven years and judgment is passed on the quality of the universities academic output to determine its allocated funds. Each academic department is required to collect published articles and books written by its scholars. In the 2008 review there were sixty-seven disciplines (over 200,000 items of scholarship) reviewed by panels of specialists (ten to twenty for each panel). Each submitted work receives a grade of one through four (four is the top grade meaning its "quality is world leading in terms of originality, significance and rigor"). The criticism of the methodology is that its roots are in the corporate not the academic world imposed on academia by politicians as the raw material of Key Performance Indicators. KPI can be manipulated to show academics are or are not providing value for taxpayers' money. The system has shifted the balance of power in British universities from academics to managers not only including central administrators but also academics who are responsible for submitting work to the RAE panels. They have become hybrid academics/managers and like in business they must please those who control the purse strings. Academics writing articles and books must pass the scrutiny of the RAE line manager who focuses on the influence of the journal, the number of citations of the text, the amount of pages written, or the journal's publisher. In the natural and social sciences pressure exists to publish journal articles rather than books. Journal articles and monographs work well because they can be completed and peer-reviewed in advance of the RAE deadline. Peer approval in prestigious journals impresses RAE panelists.[15]

In the United States higher public education is the responsibility of each state and private universities are not controlled by either the state or federal governments thus ensuring that no authority (like the HEFCE) will exercise monopolistic powers over research funding any time soon. Tenured American professors with lifetime employment (nearly fifty percent of higher education teachers) have the confidence to stand up to university managers. The tenured professors do relatively little teaching, especially on the undergraduate level, and are generally left peacefully to do their research. The majority of American institutions of higher education are public (state and two-year junior and community colleges) and have experienced a decline in funding from state and local governments even before the 'Great Recession of 2008.' University managers are acting more and more like their corporate counterparts treating their departments as 'cost centers and revenue production units.' Texas A&M University of College Station, Texas is seemly the first of a 'teaching factory' where each faculty member is given a 'profit and loss account.' The account is based on whether the teaching revenues brought in by the faculty member's 'semester credit hours' is greater or lesser than the cost of the faculty member's salary and benefits. A professor's research and publication record is not included in the profit and loss calculations.[16]

At the classroom level the mass production of teaching is about marketing and how many bodies can be processed. Teaching loads are increasing as faculty numbers and remuneration are being strictly controlled. The growth of the contingent academic workforce . . . i.e., non-tenured and without secure benefits has grown significantly over the last three decades. This workforce includes adjuncts, doctoral students, and academics employed on short-term contracts many of whom work part-time with little job security and no benefits. Of all the new full-time faculty hired between 1993 and 2003 those employed on short-term contracts without the prospect of tenure increased from fifty percent to almost fifty-nine percent. Between 1976 and 2005 the full-time contingent academic workforce grew by 223 percent, the part-time contingent workforce grew by 214 percent, while the tenured and tenure-track workforce grew by 17 percent. This trend gives academic managers a much more flexible, low-cost workforce that can be hired and fired at will. Depending on the market (i.e., number of students and credit hours) the workforce can be made to work longer or shorter hours. The workforce has little or no power to demand higher compensation. It clearly appears, especially since the 'Great Recession of 2008' that the academic world is quickly being dominated by a contingent workforce in a post-tenured academic world. Academic disciplines may be replaced by 'client services' as the organizing principle of 'instructional delivery' (i.e., teaching). Academic life is becoming much more corporate with the faculty serving as managed professionals and a 'renegotiation' of the social contract between faculty and the institution with less emphasis on academic values.[17]

REFERENCES FOR ECONOMICS AND BUSINESS

1. Cohen, Daniel; translated by Jessica B. Baker (2006). *Globalization and Its Enemies.* Cambridge: MIT Press.
2. Berger, Suzanne (2006). *How We Compete: What Companies Around the World are Doing to Make It in Today's Global Economy.* New York: Doubleday.
3. *The Economist* (March 9, 2006). "Power to the People ... Iqbal Quadir pioneered wider access to mobile phones in Bangladesh. Can he do the same for electricity and clean water?"
4. *The Economist* (March 9, 2006). "Dancing with Google's spiders ... search engines: Google is engaged in elaborate dance with firms determined to keep their websites high up in its rankings."
5. *The Economist* (March 9, 2006). "Listening to the internet ... Internet trends: Companies are eavesdropping on online discussion forums to find out what their customers really think about them."
6. *The Economist* (March 9, 2006). "Signs of the times ... Advertising technology: Huge video screens that bombard people with ads while they shop offer an attractive new outlet for advertisers."
7. *The Economist* (March 9, 2006). "Television's next big shift ... Consumer electronics: Will 'placeshifting,' which

lets you watch your TV from anywhere, be as disruptive as timeshifting?"

8. *The Economist* (February 25, 2010). "A special report on managing information . . . a different game . . . Information is transforming traditional businesses."

9. *The Economist* (February 25, 2010). "Clicking for gold . . . How internet companies profit from data on the web."

10. *The Economist* (November 4, 2010). "A special report on smart systems . . . augmented business . . . Smart systems will disrupt lots of industries, and perhaps the entire economy."

11. Coyle, Diane (1998). *Weightless World: Strategies for Managing the Digital Economy.* Boston: MIT Press.

12. *The Economist* (November 4, 2010). "A special report on smart systems . . . The IT paydirt Who will clean up?"

13. Higher Education Funding Council for England (HEFCE) (2008). "Strategic Plan, 2006-2011" www.hefce.ac.uk/pubs/hefce/2008/08_15/

14. Kaplan, Robert; Norton, David (February 1992) (September-October 1993). "The Balanced Scorecard: Measures that Drive Performance," "Putting the Balanced Scorecard to Work" *Harvard Business Review.*

15. Head, Simon (January 13, 2011). "The Grim Threat to British Universities" *The New York Review of Books.*

16. Schuster, Jack; Finkelstein, Martin (2010). *The American Faculty: The Restructuring of Academic Work and Careers.* Baltimore: Johns Hopkins University Press.

17. Slaughter, Sheila; Rhoades, Gary (2010). *Academic Capitalism and the New Economy.* Baltimore: Johns Hopkins University Press.

ARTIFICIAL INTELLIGENCE

Artificial Intelligence (AI) is the intelligence of machines and the branch of computer science that aims to create it. John McCarthy coined the term in 1956. AI involves intelligence agents that perceive its environment and take action to maximize success. Intelligence is the central property of humanity that differentiates us from all other earthly life. AI posits that human intelligence can be simulated by a machine. General Intelligence (i.e., strong AI) is still among the field's long term goals. The concept of AI goes back to every major civilization in ancient times who built animated human-like statues that were believed by some to have intelligence. Pamela McCorduck argues that our fictional characters such as Mary Shelley's Frankenstein are examples of an ancient urge "to forge the gods" and present the same hopes, fears, and ethical concerns that are presented by Artificial Intelligence.[1]

The study of logic and the application of mathematical principles have led directly to the development of today's programmable digital electronic computers based on various combinations or strings of zeros and ones. If we can simulate any conceivable act of mathematical deduction why not combine with the fields of neurology, information theory, and cybernetics to build an electronic brain? Thus far in the 21st century AI is used for logistics, data mining, and medical diagnosis. The general challenges of Artificial Intelligence are sub-categorized: deduction, reasoning, and problem solving; knowledge representation; planning; learning; natural language processing; motion and manipulation; perception; social intelligence; creativity; general intelligence.

Algorithms have been developed to imitate human step-by-step reasoning, problem solving, and logical deductions. Employing concepts from probability and economics the algorithms have been successful in dealing with uncertain or incomplete information. When the problem goes over a certain size the algorithms require enormous computational resources. Humans can solve problems quickly by using intuitive judgments rather than step-by-step deductions. AI has made progress imitating this kind of 'sub-symbolic' problem solving. The 'embodied agent' (aka interface agent such as a mobile robot) approaches emphasize the importance of 'sensorimotor' skills (from Piaget's theory of cognitive development) to higher reasoning and 'neural net' (artificial neural networks are made up of interconnecting artificial neurons that are programming constructs that mimic the properties of biological neurons) research thereby simulating human and animal brains that give rise to such a skill.

Many of the world's problems require extensive knowledge to resolve. Solid research requires us to know what we know about what other people know (i.e., knowledge about knowledge). AI needs to represent: objects, properties, categories, and relations between objects; situations, events, states, and time; causes and effects; knowledge about knowledge. Ontological (the philosophy of 'being') knowledge for AI is the complete representation of what exists. If a group of people bring up amphibious animals in a conversation we typically visualize a large, predatory, awkward animal. None of these things are true about all amphibious animals. This qualification problem is based on our default reasoning ability in which there can be a huge number of exceptions. Abstract logic requires things to be true or false and almost nothing is. The amount of facts or the amount of information the average person has is huge and requires laborious ontological engineering to give a machine a complete database of commonsense knowledge. An AI goal is to have the machine understand enough concepts to be able to read sources (e.g., the Internet) so as to add to its own ontology. Human intuition is represented in our brains unconsciously and 'sub-symbolically' because what we know is not all facts or statements that can be expressed verbally. We know certain things because they or it do not 'feel' right: e.g., a group of new acquaintances or a particular restaurant.

Machine Learning

Intelligent agents must be able to plan . . . to set goals. They need to be able to predict how their actions will change the environment; they need the ability to maximize the value (utility) by choosing from available options. As the environment and world changes the plan and action may need to change based on new circumstances/situations. This is done by checking if the environment is matching the agent's choice of options . . . if it is not the agent may have to reason with uncertainty. Some goals require multi-agent planning by using cooperation and competition of many agents. Swarm intelligence is made from a population of simple agents or 'boids' interacting with one another and the environment without central control dictating behavior and leading to the emergence of 'intelligent' global behavior that is unknown to the individual agent. Natural examples are ant colonies, bird flocking, animal herding, and fish schooling.

Machine unsupervised learning is the ability to find patterns in a stream of data. Supervised learning includes both classification (what category something belongs to) and numerical regression (from a set of input/output samples attempts to discover a continuous function to generate the outputs from the inputs). Natural language processing gives machines the ability to read and understand spoken languages. It is a goal to have the machine acquire knowledge on its own by reading on the Internet. Two simple applications are information retrieval (text mining) and machine language translation. Motion and manipulation takes us to the closely related field of robotics and what any useful robot needs. If AI is to be functional and useful and not just another 'dumb' machine it needs to handle such tasks as manipulating objects as humans do so effortlessly (grasping, squeezing, dropping, placing, moving, throwing, etc.). It needs to be able to navigate (mobility) and know where it is (localization). It needs to able to map (learning what is around you) and have motion planning (figuring out how to get somewhere).

Many humans take their senses for granted . . . makers of machines do not! Machine perception is the ability to use (like humans) input from sensors (e.g., cameras, microphones, sonar, etc.) to deduce aspects of the environment and analyze what it means/represents and what actions are required. Computer vision is the ability to analyze input. Sub-challenges of perception is speech recognition, facial recognition, and object recognition.

Social intelligence is a major human intelligence that is critical to our culture and well being. Emotion and social skills are not small accomplishments for a machine. It must be able to predict the actions of others by understanding their motives and emotional states in addition to be able to model human emotions and the perceptual skills to detect emotions. For quality human-computer interaction it needs to display emotions and appear polite and sensitive. Having high levels of human social intelligence is something not every human has. Creativity is likened to artificial intuition and artificial imagination. Creativity is approached theoretically from a philosophical and psychological perspective. It is also approached pragmatically by specific system implementations that produce outputs considered creative or systems that identify, assess, and evaluate creativity.

Strong AI or general intelligence combines all the aforementioned skills and exceeds human abilities in all of them. Some researchers believe that an artificial brain or artificial consciousness may be required and some believe in the concept of AI-complete: i.e., to solve one problem, all of the problems must be solved. Machine translations for example are believed to be AI-complete: i.e., it may require strong AI to be done like a human being. Machine translations require reason, knowledge and social intelligence.[2]

In Michael Crichton's thriller "Prey" the parents of three children, Jack and Julia work in Silicon Valley for MediaTronics and Xymos respectfully. Xymos is developing nano-robots, i.e., tiny machines that can function autonomously but are programmed to work together like an army of ants. MediaTronics makes software to coordinate the actions of autonomous agents such as nano-robots giving them intelligence and flexibility. Jack loses his job and Julia becomes involved in a secret project to develop the nano-robots into a stealthy photo-reconnaissance system for the U.S. Army. She introduces living bacteria into the nano-robots so they can reproduce and evolve rapidly and increase the system's power and performance. They are re-programmed by MediaTronics software so they can learn from experience. Xymos loses its Army funding and Julia tries to convert the system into a medical diagnostic tool in which the nano-robots will be trained to enter and explore the human body to locate pathological conditions. Julia experiments with her own body and becomes chronically infected as the nano-robots become symbionts within her body and gradually take control of her. She becomes insane infecting three of her colleagues and sets her creation loose where they prey on wildlife quickly increasing their population.

As Jack slowly realizes that something is terribly wrong with Julia he confronts her and douses her with bacteriophage (lethal to the bacteria in her). Jack and a loyal friend also douse the infected colleagues and like Julia, they all collapse and die. They also destroy the nano-robots inside and outside of the laboratory with fire and explosives. The film ends with Jack at home with his children wondering if his actions were enough to kill all the critters and whether Xymos has others or is developing similar projects that will turn nightmarish.[3]

Was Crichton prophetically warning us about what lies ahead if our technological developments are allowed to continue? In his introductory chapter he writes that he intended for his fiction to be taken seriously. It is easy to demonstrate the scientific flaws of Crichton's fiction. Nano-robot's camera would have to be smaller than a red blood cell. Swarms of nano-robots chase Jack swarming through the air like bees as fast as he can run. The laws of physics do not allow very small creatures to fly fast because of the viscous drag of air or water strengthens as the creature becomes smaller. To fly through the air like bees they would have to be the same size as bees. The flying swarms are said to be solar powered but they are too small for this to occur . . . just not enough energy. The point of the story, however is not about science and physics, it is about us . . . human beings.

Julia is a skilled scientist whose family and employer are in a deep financial crisis. Xymos could go bankrupt causing Julia to lose her employment. She takes the scary step of pushing ahead with risky technology in a high stakes game of survival. "Prey" is giving us an important message about biotechnology in the twenty-first century and it is not about any particular autonomous agents. It is about the dangers of humanity's growing knowledge and understanding of life's basic processes. Irresponsible use of this knowledge and understanding leads to death. Is the world listening?

Precautionary Principle

What should our response be to dangers that are poorly understood and unproven? There are two opposing points of view about our response. The 'precautionary principle' posits that whenever there is a risk of a major disaster no action should be allowed that increases the risk even if it promises substantial benefits. No balancing of benefits against risks is to be allowed. The costs of prohibition to avoid the risk of a major disaster does not matter. The opposing view holds that 'risks are unavoidable' and no course of action or inaction will eliminate them. A balancing of risks against benefits and costs is the best course. How do we calculate the cost of human life? Insurance companies and our legal systems do it all the time. Will scientists and corporations calculate risks like Mao Tse Tung's famous comment that China with the world's largest population could survive a nuclear war? Bill Joy, co-founder and chief scientist of Sun Microsystems said in "Wired" magazine (April 2000):

> "Our most powerful 21st century technologies . . . robotics, genetic engineering, and nanotech . . . are threatening to make humans an endangered species. The 21st-century technologies – genetics, nanotechnology, and robotics (GNR) – are so

powerful that they can spawn whole new classes of accidents and abuses. Most dangerously, for the first time, these accidents and abuses are widely within the reach of individuals or small groups. They will not require large facilities or rare raw materials. Knowledge alone will enable the use of them. Thus we have the possibility not just of weapons of mass destruction but of knowledge-enabled mass destruction (KMD), this destructiveness hugely amplified by the power of self-replication. I think it is no exaggeration to say we are on the cusp of the further perfection of extreme evil, an evil whose possibility spreads well beyond that which weapons of mass destruction bequeathed to the nation-states, on to a surprising and terrible empowerment of extreme individuals."

Bill Joy became a 'precautionary view' spokesman writing his warning eighteen months prior to 9-11-01. At the annual meeting of the World Economic Forum held in Davos, Switzerland in 2001 Joy quoted Eric Drexler, the chief prophet of nanotechnology and originator of the Foresight Institute to promote benign uses of nanotechnology and warn against its dangerous uses.

"Tough omnivorous [synthetic] 'bacteria' could out-compete real bacteria: They could spread like blowing pollen, replicate swiftly, and reduce the biosphere to dust in a matter of days. Dangerous replicators could easily be too tough, small, and rapidly spreading to stop – at least if we make no preparation. We have trouble enough controlling viruses and fruit flies . . . . We cannot afford certain kinds of accidents with replicating assemblers. The idea of nanotechnology is to build machines on a tiny scale that are as capable as living cells, but made of different materials so that they are more rugged and more versatile. One kind of nano-machine is the assembler, which is a tiny factory that can manufacture other machines, including replicas of itself."

A replicating assembler would be a tool of immense power for good or for evil. Nothing resembling an assembler has yet emerged. The most useful products of nanotechnology so far are computer chips. They have no capacity for replicating either themselves or anything else.

Physicist Freeman Dyson writing in the New York Review of Books in 2003 offered some first steps to avoid a disaster with our new technologies:

1. Have scientists and technologists (and corporate leaders as well) take a vow, along the lines of the Hippocratic Oath, to avoid work on potential and actual weapons of mass destruction;
2. Create an international body to publicly examine the dangers and ethical issues of new technology;
3. Use stricter notions of liability, forcing companies to take responsibility for consequences through a private-sector mechanism – insurance;

4. Internationalize control of knowledge and technologies that have great potential but are judged too dangerous to be made commercially available;

5. Relinquish pursuit of that knowledge and development of those technologies so dangerous that we judge it better that they never be available.

Dyson who debated Joy at the World Economic Forum in Switzerland agreed with the described dangers but disagreed with his remedies. Beginning with the successful history of the international community to regulate and prohibit dangerous technologies such as biological weapons and gene-splicing experiments Dyson cited the call for a moratorium on gene-splicing experiments in 1975 by the world's two leading biologists Maxine Singer and Paul Berg. Because of the public health dangers biologists worldwide quickly agreed to the moratorium and experiments stopped for ten months. Two international conferences were held during the moratorium and guidelines were established to continue certain experiments involving various degrees of risk; the guidelines were adopted voluntarily and have been observed ever since with changes made in response to new discoveries. No new serious health hazards have arisen from gene-splicing experiments since 1975.

Biological weapons have a different and more disturbing history. Compared to nuclear weapon development and procurement, biological weapons were developed and stockpiled much more quietly. Richard Nixon declared in 1969 that the United States was stopping its production and destroying its stockpile of biological weapons unilaterally.

Britain quickly did the same with its biological weapons and in 1972 an international convention was signed by the United States, the United Kingdom, and the USSR imposing a permanent prohibition of biological weapons on all three countries. Many other countries subsequently signed the convention. The former Soviet Union violated the Convention of 1972 very significantly. Post Soviet Union, Russia declared its adherence to the convention but has never provided convincing evidence that the program is not continuing as many of their research and production centers remain cloaked in secrecy. It is very feasible that biological weapons continue to exist in Russia and other countries. The dangers of biological weapons have not been eliminated by the Biological Weapons Convention of 1972 but it has been reduced.

There is very different opinions about the remedies for the dangers of these bio-technologies. Bill Joy wants to: "internationalize control of knowledge and relinquish pursuit of that knowledge." Freeman Dyson is opposed to the censorship of scientific inquiry either by national or international authorities. Some argue that the risks of modern biotechnology are historically unparalleled since the dangers of letting new living creatures loose may be irreversible. Dyson uses the analogy of printed books in the seventeenth-century (when fear of moral contagion by soul-corrupting books was a grave concern) to the twenty-first-century fear of physical contagion by pathogenic microbes. Neither the seventeenth-century or the twenty-first-century fear was groundless or unreasonable. The English Parliament during the religious wars of the seventeenth-century (thirty-year war) saw books not only corrupting souls but also responsible for killing people. Letting books go free into the world were regarded as potentially lethal as well as irreversible. Poet John Milton argued that the risks of the unlicensed printing of books must be accepted. Dyson uses the same argument for biological experiments since not allowing either books or experiments to ever see the light of day (through censorship and prohibition) is just not acceptable to the standard bearers of individual freedoms.

Dyson relates to the effectiveness Nevil Shute's 1957 best selling book "On the Beach" that was made into a very successful film, had on public opinion about the dangers of nuclear war. As a scientist who pointed out the flaws in Crichton's book "Prey" he does the same with "On the Beach." He claims that the book and the film created an enduring 'myth' about nuclear war as a silent inexorable death from which there is no escape, with radioactive cobalt sweeping slowly down the sky from the Northern and Southern Hemisphere. The people of Australia, after us northerners are all dead, are the only remains of humanity. The Australian government provides euthanasia pills for all citizens to take when the radiation sickness becomes intolerable. There is no hope of survival or plans to go underground until the cobalt decays. Shute described humanity graciously accepting its own extinction. Dyson correctly points out that the type of radiological warfare Shute uses to cause the end of humanity is technically flawed in many ways (radioactive cobalt would not increase lethality of hydrogen bombs; fallout would not descend uniformly; people could go underground to survive; no country had enough bombs in 1957 to lethally radiate the entire planet). It is generally agreed that a nuclear holocaust may not eliminate all human life and certainly not all earthly life but it certainly would not create a very pleasant life for whoever survives.[4] The 1995 film "Twelve Monkeys" was the biological equivalent to "On the Beach" in which a crazed scientist releases deadly biological contagious gases at an international airport quickly causing worldwide illness and death. The survivors are forced to exist underground and send a convict back in time to hopefully prevent the release of the gases and thereby cause a different future . . . he fails his mission.

Hackers

The word 'hacker' generally refers to someone who enjoys tinkering with technology, exploring its boundaries and getting it to do unexpected or unintended tricks. Many people see 'hackers' as people who break into computers for nefarious reasons. Technological tinkering does not need to be confined to computers. Getting cars, cameras, and household appliances to do things not originally intended is becoming more popular with 'hackers.' Modifying cars has become more complex with a much greater emphasis on electronics; e.g., replacing one or two microchips can provide more power and less fuel efficiency. Hackers are hacking electronic games such as Nintendo to become tuning tools for cars that allow running engine reconfigurations. The built-in diagnostic systems of hybrid and electric cars are being hacked for various reasons including improving the car's range on each charge. Some of the hybrids are being hacked to use the electrical power more frequently thus changing the gasoline consumption and driving range per charge. Some hybrids are being outfitted with larger lithium-ion battery packs that can be adjusted to extend the car's electric-only range.

As the complexity of consumer-electronic devices increases they become more ripe for hacker modification. The TiVo personal video recorder that initiated a dramatic shift in viewing habits allowed for a wide variety of modifications: different colors; installing a larger hard disk to increase recording capacity that evolved into new software development allowing for Internet remote control. TiVo users downloaded programs to their laptops and/or transferred them to DVD. Instructions on how to do this is free on the Internet. Since TiVo runs on Linux software, it was utilized to run other devices not based on Linux thus allowing discovery of proprietary hardware secrets. For example Apple's ipod music player can record high-quality audio which was not the intention of Apple to allow. Ipods were also manipulated to run games and display pictures. Some businesses like Microsoft will sell equipment at a loss in order to make profit on each game or app sold and pass on part of the profit to the hardware maker (e.g., game consoles). By hacking the console (e.g., Microsoft's Xbox) users did not have to purchase the games because they transformed it into a low-cost, high performance media-playback system (Microsoft beefed up its security measures with the new Xbox 360 console). Compromising a company's business model was also evident with the pharmacy chain CVS that sold low-cost 'disposable' digital cameras. The cameras were used once and returned to the shop where they were processed and returned to the user on a CD or DVD; the cameras were then reset and resold. Hackers learned how to access and reuse the camera themselves breaking down the original intent of the business model. Some companies are embracing customers who like to tinker (i.e., hackers) because they are willing to spend their own time and effort to improve a product. The company iRobot that produces the Roomba robot vacuum-cleaner is one such company. Their vacuum-cleaner comes with an external data connector that allows customers to reprogram the device.[5]

IBM's Watson

Since 2007 IBM has been developing the world's most advanced 'question answering' machine. A machine that can answer any question posed in 'natural language' (i.e., common human elocution). Unlike Google and Bing search engines, IBM's machine not only finds the sources where the answer may be found but exactly what the correct answer is. This capability is at the core of artificial intelligence because it allows us to converse with machines . . . something way beyond typing 'keywords;' A machine that became a champion on the popular television show "Jeopardy." The machine is called 'Watson' and the producers of "Jeopardy" allowed Watson to compete on three shows in February 2011 against the most successful past champions (Ken Jennings and Brad Rutter). Jennings and Rutter were no match for Watson and it won by a large margin. Like the human competitors Watson is not hooked up to the Internet.

It draws its answers from its vast database (a roomful of servers); like humans draw their answers from the database in their brain. In the "Jeopardy" shows and in trials Watson did exceedingly well and with its machine-synthesized voice it sounds similar to the "War Games" movie computer that almost destroyed the world by starting a nuclear war. The contestants Watson competed against (in the trials and on live television) referred to it as 'he' and made references to "Skynet," the computer system in the "Terminator" movies that achieved consciousness and decided that all humans must be eliminated.

IBM has a successful history of having their machines compete with humans. In 1997 IBM supercomputer 'Deep Blue' beat chess grandmaster Garry Kasparov. The feat had wide-world publicity but did not produce any direct profit for the company. Watson could be the first of a valuable profit making product for any number of industries: legal firms that need to swiftly sift through case law to find precedent and citations; help-desk workers that need to find answers for customers from huge databases of product information.

Since the dawn of the 21st century algorithms have increasingly become more complex, more and more able to analyze huge piles of information, and statistically determine what types of words/phrases/subjects are associated with any subject. With the explosion of the Internet, access to data, storage, retrieval, and lower cost machine capabilities concurrently exploded. Watson thinks in probabilities and uses more than a hundred algorithms at the same time to analyze a question in different ways, generating hundreds of possible solutions. Another set of algorithms ranks these answers according to plausibility; for example, if dozens of algorithms working in different directions all arrive at the same answer, it's more likely to be the right one.

IBM plans to begin selling commercial versions of Watson in the near future. A medical version of Watson is envisioned to give speedy answers in the emergency room. A virtual call center is also envisioned where Watson-like-units could talk directly to customers and be the first line of defense . . . the first responder. Any industry where time is critical and advanced state-of-the-art information is needed by decision-makers can be a potential IBM customer for Watsons. Today Watson servers fill up a room but if the past is any predictor of the future it should quickly end up in laptop size or smaller. Many experts are already thinking of offshoots to Watson such as a 'contradiction engine' that provides evidence opposite of a word/phrase/thought/fact.

The plans for Watson's commercial equivalents has some obstacles to overcome. Many Wall Street companies rely on 'millisecond trading' machines that are now regarded by many as the culprit in the irrational hour-long stock market plunge in May, 2010. Physicians clearly will need some time to accept Dr. Watson in a crisis. Watsons can save money for firms but labor unions have traditionally deplored labor-saving advances. With increasing computerized telephone services many people continue to want to talk to a fellow human being. Watson creators believe the machine may be simulating some of the workings of the human brain, especially in regard to processing language. Our brains are highly 'parallel' using different parts at the same time whenever we talk or listen to words. Watson mimics this 'parallel' approach by looking for answers in thousands of different ways. Watson, like us, does not come up with the answer to a question as much as make an educated guess, based on similarities to things it has been exposed to.[6]

Google Cars Drive Themselves, in Traffic

Using artificial intelligence software Google has developed a car that can sense anything near it and mimic the decisions made by a human driver. Seven test cars have driven 1,000 miles without human intervention and more than 140,000 miles with occasional human control. Technologists believe that autonomous cars will transform society as profoundly as the Internet because robots react faster, have 360 degree perception, and do not get sleepy, intoxicated, or distracted. The capacity of our roads could increase because greater safety and reaction time means cars can drive closer together. Robots are less likely to crash so the cars could be built lighter, reducing fuel consumption. With Global Positioning Software (GPS) technology and detailed files on every road and their speed limits the robot cars are very apt at keeping within the law. The cars can be programmed for different driving personalities from cautious to aggressive . . . e.g., at a four-way stop intersection the cautious would allow the other drivers to go first whereas the aggressive would be more likely to go first. Human passengers can gain control of the car at any time by pushing a button or touching the steering wheel or brake. The cars each have a sensor mounted in the center of the vehicle's roof known as Lidar, or Light Detection and Ranging to provide a continuous updated three-dimensional map of the world at centimeter accuracy extending for over two-hundred feet around the car. Lidar is supplemented with four standard automotive radars with even greater range (three in front one in the rear). Next to the rear-view mirror is a high resolution video camera to detect street lights and moving obstacles such as pedestrians and bicycles. In addition to the latest in GPS technology the vehicle also is equipped with an inertial motion sensor. The cars drive autonomously over selected routes recording changes as they occur and update the map.[7]

Aiming to Learn as We Do, a Machine Teaches Itself

Early 21st century machines can do tasks that are clearly defined such as win at chess and predict the weather. The challenge has been when issues are nuanced or ambiguous. Humans learn over their lifetime from experience especially when it comes to understanding the meaning of language and unraveling semantics and nuances. Carnegie Mellon University is working on a computer system to master semantics and learn more like a human. It has been given knowledge in various categories and combs the Web 24/7 with a mission to teach itself. By scanning hundreds of millions of Web pages for text patterns used to learn facts the "Never-Ending Language Learning System" (NELL) has grouped its knowledge into semantic categories: cities, companies, sports teams, actors, universities, plants, and almost three-hundred others.

NELL learns facts that are also relations (hundreds) between members of two or more categories: e.g., Barry

Bonds is a baseball player. San Francisco Giants is a baseball team. NELL infers that Barry Bonds plays for the San Francisco Giants. 'Plays for' is a relation with almost three hundred kinds of relations. The number of categories and relations are steadily expanding as NELL's knowledge base expands. NELL's learning algorithms are refined as its pool of facts grows thus allowing it to find facts more accurately and efficiently over time. NELL is one project in an expanding field of research to enable machines to better understand the meaning of language and mimic human understanding. Progress is occurring rapidly with universities, government labs, Google, Microsoft, I.B.M., and elsewhere pursuing breakthroughs along somewhat differing paths.

NELL differs from I.B.M.'s 'Watson' (a question answering machine) and 'Google Squared' (a research project that grasps semantic categories as it finds and presents information) in that it is not passive but highly automated and a continuous learner. NELL like a human exercises curiosity on its own. Machine understanding of language has many potential applications: smarter searching, personal assistants for questions about health, education, travel, and shopping. NELL uses tools that extract and classify text phrases, look for patterns and correlations, and learn rules. By deploying a hierarchy of rules it can resolve ambiguity. For example "I climbed XXX" very often occurs with a mountain but NELL has learned that "I climbed stairs" belongs to the category of "building part." It self corrects when it learns more and has more information and context to choose from. By receiving human attention to correct errors, NELL has very able partners/parents as it learns more and more about just about everything. For example when NELL associated "Internet" cookies with baked goods, human intervention gave NELL a whole new lesson in word use and category. The ideal for NELL's innovators is to produce a computer system that can learn continuously with no need for human assistance.[8]

A Soft Spot for Circuitry: Robot Companions

Paro is a robot modeled after a baby harp seal and manufactured in Japan. It trills and paddles when petted, blinks when the lights go up, opens its eyes at loud noises, and yelps when handled roughly or held upside down. Under its artificial white fur are microprocessors and hidden sensors for sound, light, temperature, and touch. It perks up at the sound of its name, praise, and words it hears most frequently. The name Paro is derived from the first sounds of the words 'personal robot' and is one of a handful of devices designed to soothe, support, and keep us company. For recovering drug addicts there is a wearable sensor that senses drug cravings and sends tough love text messages. A talking robotic head can be modeled to the personality of the buyer's choice. A dieter's robot sits on the kitchen counter and offers encouragement after calculating calories and exercise.

In its earliest manifestations 'machines as companions' take advantage of the innate soft spot many people have for objects that seem to care . . . or people who need someone to care for them. Robots are appearing in nursing homes, schools, and some living rooms thus moving science fiction into reality. The debate over what human responsibilities machines should and should not be allowed to undertake is well underway. The nursing home patients that have taken Paro as a pet and a companion are not complaining and neither are their administrators and nurses who see Paro's very significant therapeutic value. Unlike live pets, there is no allergy problem, it need not be fed or cleaned up after, it does not bite. In some cases it offers an alternative to medications. Critics of Paro and its ilk argue "who among us will eventually be deserving enough to deserve real people?" The growing popularity of robots, even in their early stage of development, is that robots, unlike people do not feign interest in another or abruptly switch off their affections.

Timothy Hornyak, author of "Loving the Machine" argues that humanity has to learn how to deal with a new range of synthetic emotions; i.e., emanating from a manufactured object. His book is about Japanese robots, where the world's most rapidly aging population is accepting robotic care more and more. He believes that our technology is getting ahead of our psychology. Japan started its artificial pet industry with Aibo, the metallic dog and Furby the talking pet. Paro is much better at emotional bonding and less toy like . . . its successors are sure to be even more advanced. These robots may not replace our best friends, for those of us fortunate to have a best friend(s), but as their manufacturing costs continue to drop along with their selling price, they are sure to gain in popularity.

Paro has started a new process called 'robot therapy' in which elderly patients can get together in groups with one or more Paros and take turns grooming, petting, and crooning with the robotic seal. Visitors from a retirement home gather weekly at the Veterans Affairs Medical Center in Washington with two Paros. Staff have noticed that the robots stimulate human conversations rather than replacing it. Several patients whose mental faculties are intact have made special visits to Paro between their weekly sessions; they know it is not real but it brings out natural feelings. When something responds to us our emotions are hardwired to respond to it . . . alive or not.

The diet coach robot Autom developed at the M.I.T. Media Lab makes eye contact critical to its appeal with a female voice since most people view women as supportive and helpful. Autom blends dispassion with personal attention making it appealing and helpful to serious dieters; it always has a plan B if the user does not like or follow its initial advice. System improvements are being developed for drug addicts with sensors to detect physiological data correlated to cravings and wireless signals to send supportive and helpful text messages. With improved algorithms and GPS units the system could select alternate routes when approaching places that hold particular temptations and/or play a motivational song or show pictures of the user's children. The systems are designed to be there for the user and it works best if the user see it as a trustworthy companion.[9]

Students, Meet Your New Teacher, Mr. Robot

Computer scientists around the world are developing robots as highly programmed machines that can engage people and teach them simple skills, including household tasks, vocabulary, playing, elementary imitation, and taking turns. The most advanced models are fully autonomous guided by software that make them engaging enough to rival people at some teaching tasks. They can track motion and recognize speech. In highly repetitive subjects like language or therapies these machines should begin to learn as they teach becoming a very patient, highly informed instructor. They could be very effective to treat people with developmental disabilities. Proponents see them as supplemental adjuncts to classroom learning. Detractors see children viewing them as the instructors and masters. The University of California at San Diego uses its teaching robot RUBI to teach preschool children languages and has enabled its students to score significantly higher on tests, compared to less interactive learning (e.g., tapes). Students with human teachers and students with RUBI scored equally well on tests. Researchers have found that including features on the robots that make them most convincingly real as a social partner, a helper, a teacher is most beneficial to being accepted and is part of a new field called 'affective computing.' The key is in the machines behavior not how much it looks human: e.g., if RUBI reacts too fast or too slow it throws off the child's expression or comment, but if it reacts within 1.5 seconds, the child and machine became more synchronized. Bobbing or shaking in the same rhythm as the child can quickly engage even the most fearful. Once synchronous behavior is in play then social behaviors can be addressed like eye contact, joint attention, and turn taking. Like good teachers robots must learn from the students when a lesson is working and when it is not and make adjustments accordingly. Researchers are searching for the foundation of human learning or at least its artificial intelligent equivalent so as to make teaching robots far more effect and responsive to individual needs.[10]

Scientists Worry Machines May Outsmart Man

Computer scientists are debating whether or not there should be limits on research that might lead to a loss of human control with its machines. Computer-based systems are carrying an increasingly greater share of society's workload from waging war to having telephone conversations. Further advances could create profound social disruptions and possibly have dangerous consequences. Medical systems are interacting with patients to simulate empathy and computer worms/viruses that are defying extermination.

Meeting February 25, 2009 at the Asilomar Conference Grounds on Monterey Bay in California leading computer scientists, artificial intelligence researchers, and roboticists did not think that 'Hal' (the computer that took over the spaceship in the movie "2001: A Space Odyssey") is around the corner. They are concerned about destroying an ever widening range of jobs and forcing us to live with a plethora of human-mimicking-machines. They are concerned about the autonomous killer robots that are already here and more exploitation of A.I. by criminals. Criminals could use masquerade speech synthesis systems as a person or mine personal information for their own nefarious schemes. The conference organized by the Association for the Advancement of Artificial Intelligence choose Asilomar because of its exalted place in science history. In 1975 the world's leading biologists met there to discuss the new ability to reshape life by swapping genetic material among organisms. Certain experiments were halted and guidelines issued for recombinant DNA research enabled experimentation to continue. In 2009 the concern was 'super-intelligent' machines and artificial intelligent systems run amok. The rising voice of the 'technorati' and people very concerned about the rise of intelligent machines has not yet produced similar guidelines like those in 1975 for genetic research. Technologists like Raymond Kurzweil are providing almost religious visions similar to the concept of Rapture.[11]

The Coming Super-brain

The rapid rise of Artificial Intelligence (A.I.) is turning the spotlight onto the ominous; will machine intelligence surpass human intelligence and how soon? Vernor Vinge, computer scientist and science fiction writer wrote about ultra-smart computers (i.e., those with intelligence greater than human intelligence) and dubbed the term 'The Singularity' in a 1993 paper. For Singulatarians, A.I. refers to machines that will be both self-aware and superhuman in intelligence. They will be capable of designing better computers and robots faster than people. This shift will lead to a vast technological acceleration and all kinds of improvements. Singularity is no longer just a science fiction notion . . . many highly respected scientists and businesspeople embrace the idea of exponential technological change as explained in "Moore's Law" (Gordon Moore is a co-founder of the chip maker

Intel). In 1965 Dr. Moore described the repeated doubling of transistors on silicon chips with each new generation leading to rapid acceleration in the power of computing. Moore's Law is a description of the rate of industrial change. Artificial Intelligence pioneer Raymond Kurzweil in his 2005 book "The Singularity Is Near: When Humans Transcend Biology" expanded Moore's Law beyond processing power to post-human evolution which he said would occur in 2045. Unimaginable computing power combined with cyborg humans is foreseen as the next major point of our species evolution with unpredictable consequences. Dr. Kurzweil co-founded 'Singularity University' supported by Google with the goal to "assemble, educate, and inspire a cadre of leaders who strive to understand and facilitate the development of exponentially advancing technologies and apply, focus and guide these tools to address humanity's grand challenges." Dr. Kurzweil visions go beyond developing superhuman machines . . . he envisions 'uploading,' or the concept that our brain's contents and thought processes can be translated into a computing environment, making a form of immortality possible. Other scientists are describing his visions with super machines as a new form of religion. Kevin Kelly the founding editor of the magazine "Wired" forecasts the emergence of a global brain via the world's interconnected computers coordinating toward a global intelligence. William Joy, a co-founder of Sun Mircrosystems believes that humans are more likely to destroy themselves with their technology than create some utopia assisted by ultra-machines. Mr. Joy simply believes a catastrophe created by mankind is more likely to happen than an utopian society.[12]

Computers That See You and Keep Watch Over You

Prisons are utilizing cameras and artificial intelligence software to recognize faces, gestures, and patterns of group behavior in order to warn of potential or real trouble. Unlike guards watching monitors the machines are not distracted or forget to pass information onto other shifts. Machines do not blink, forget, or sleep. Prisons are not alone using high-resolution low-cost cameras . . . they are in smartphones and laptop computers. In hospitals a computer-vision system can remind physicians and nurses to wash their hands or warn of patients in danger and even read vital signs by reading a person's face. These systems can help marketers tailor what they offer based on facial expressions of online shoppers. They can be used in shopping malls, schoolyards, subway platforms, office complexes, and stadiums. Since machines are able to observe and understand us better it is both helpful and alarming. Google, facing privacy complaints eliminated homes in their 'Street View' service upon request and decided not to include facial recognition in 'Goggles' which allows users to take a picture on their smartphone and search the Internet for matching images (eliminated possibility of taking photos of people without their approval and finding personal information on them). Nevertheless computer vision is moving into the mainstream: Homeland Security, identify lost children, help locate eloped Alzheimer patients. It is projected that law enforcement, national security, and military operations will become more and more reliant on observant machines. Millions of people currently are using computer vision for online photo-sharing services using face recognition. A user puts a name to a face and the service finds matches in other photographs. Microsoft 'Kinect' is a step in the future of computing. Included in Microsoft's Xbox 360 gaming console it understands voice commands and recognizes people and gestures so players can control the game with waves of their hand to make their on-screen avatars run, jump, swing, and dance. Kinect is seen as the future world where technology sees you so you do not have to understand it.

Facial recognition software can be used in store kiosks, or with Webcams. Major retailers are interested in what expressions show product interest and possible purchase giving valuable information of product location and display. Online dating services are analyzing users' expressions in search of 'trigger words' in personal profiles that people find appealing or off-putting. By tracking movements on a few dozen facial points facial recognition software can also be used to measure reaction to movies or their trailers: the critical facial points are mostly along the eyes, eyebrows, nose, and lip perimeters. The software goes beyond human facial communication; e.g., it records precisely if a person's smiles are symmetrical (signaling amusement, not embarrassment) and not smirks. Because of Webcam's small size people in general are not bothered by it. The traditional method of survey-and-questionnaires for the likes and dislikes of a newly released movie cannot measure audience response with scene-by-scene granularity as facial recognition software can. The new technology allows producers to eliminate certain scenes or a particular character based on a solid database of viewer response.[13]

At the World Science Festival in New York, May 2008 inventor and futurist Dr. Ray Kurzweil made some positive predictions for our future with machines. He believes that before 2018 a drug will be developed that allows one to eat whatever they want without gaining weight. In regard to solar power he forecasts that nano-engineering will make exponential progress with solar power so as to be cost-competitive with fossil fuels and that before 2028 all our energy will come from clean sources. Life expectancy by 2023 will continue to rise every year even faster than we are aging. Kurzweil makes his prediction using the 'Law of Accelerating Returns,' a concept he illustrated at the festival with the history of his own inventions for people with blindness. His device that could scan books and read them aloud was rolled out in 1976. By 1990 he predicted that people with blindness would be able to read anything, anywhere with a handheld device early in the 21st century. At the festival he pulled out a new device, the size of a cell phone that read aloud when pointed at any particular text. He predicted the incredible growth of the Internet and a computer chess champion; Deep Blue's computer became a chess champion in 1997.

Kurzweil believes that certain technological aspects follow predicable growth. The first electromagnetic machines in the early 1900's doubled their power every three years. By 1950 they were doubling every two years; the rate that inspired Moore's Law. Now it takes only about one year. He predicts the fields of biology, medicine, energy and others being revolutionized by information technology. Nanotechnology will not be alone in exponential progress. He believes we will be adding computers to our brains and building machines as smart as ourselves by the 2020's.

Other neuroscientists do not share Kurzweil's confidence since our brains evolved so haphazardly that to reverse-engineer it might prove much too difficult. Many experts do believe that man can create conscious, intelligent beings. Kurzweil acknowledges the brain's complexity but believes that exponential upward curves are deceptively gradual at first. "Scientists imagine they'll keep working at the present pace." They make linear extrapolations from the past. When it took years to sequence the first 1 percent of the human genome, they worried they'd never finish, but they were right on schedule for an exponential curve. If you reach 1 percent and keep doubling your growth every year, you'll hit 100 percent in just seven years."[14]

Hans Moravec, the founder of Carnegie Mellon University's Robotics Institute sees a strong parallel between the evolution of robot intelligence and the biological intelligence that has preceded it: "The largest nervous systems doubled in size about every fifteen million years since the Cambrian explosion 550 million years ago. Robot controllers double in complexity (processing power) every year or two. They are now at the lower range of vertebrate complexity, but should catch up with us within a half a century." By Moravec's calculations no human task, physical or intellectual, will be beyond the scope of robots by the 2050's. He correctly predicted that broadly-capable 'universal robots' (servants) with mental power and inflexible behavior analogous to small reptiles will emerge in 2010 (Japan's Robovie II, Aprilpoco, Telenoid, Service Robots, Flipper-babies, Robot Chef, Toyota's Violin Playing

Robot). He forecasts that by 2015 robots will host programs for several tasks: Utility robots with manipulator arms will follow single-purpose home robots and their narrow inflexible competences will be comparable to the skills of a frog. Lizard-scale minded robots will emerge in 2020 with human-scale 'universal' robots running programs for most simple chores. Robot competence in 2030 will compare to large mammals and in the decades following mammal-like brainpower and cognitive ability will be developed; like humans they will learn from experience and adapt accordingly. They will evolve into thinking like small primates and maintain physical, cultural, and psychological models of their world to mentally rehearse and optimize tasks before physically performing them. The fourth generation will be humanlike abstracting and reasoning from the world model.[15]

REFERENCES FOR ARTIFICIAL INTELLIGENCE

1. McCorduck, Pamela (2004). *Machines Who Think: A Personal Inquiry into the History and Prospects of Artificial Intelligence.* Natick, Mass: A.K. Peters.
2. McCarthy, John (November 12, 2007). *What is Artificial Intelligence?* Stanford University Computer Science Department.
3. Crichton, Michael (2003). *Prey.* Harper Collins: U.S.A.
4. Dyson, Freeman (February 13, 2003) "The Future Needs Us!" *The New York Review of Books.*
5. *The Economist* (March 9, 2006). "Hackers go Home . . . consumer technology: Technological tinkering, or

hacking, is not limited to computers. Cars, cameras and vacuum-cleaners can be hacked too."

6. Thompson, Clive (June 16, 2010). "What is I.B.M.'s Watson?" *New York Times.*
7. Markoff, John (October 9, 2010). "Google Cars Drive Themselves, in Traffic" *New York Times.*
8. Lohr, Steve (October 4, 2010). "Aiming to Learn as We Do, a Machine Teaches Itself" *New York Times.*
9. Harmon, Amy (July 4, 2010). "A Soft Spot for Circuitry" *New York Times.*
10. Carey, Benedict; Markoff, John (July 10, 2010) "Students, Meet You New Teacher, Mr. Robot" *New York Times.*
11. Markoff, John (July 25, 2009). "Scientists Worry Machines May Outsmart Man" *New York Times.*
12. Markoff, John (May 23, 2009). "The Coming Superbrain" *New York Times.*
13. Lohr, Steve (January 1, 2011). "Computers That See You and Keep Watch Over You" *New York Times.*
14. Tierney, John (June 3, 2008). "Findings . . ."The Future Is Now? Pretty Soon, at Least" *New York Times.*
15. *The Daily Galaxy . . . Great Discoveries Channel* (March 26, 2008) "Is Robot Evolution Mirroring the Evolution of Life?"

THE INDUSTRIAL REVOLUTION AND THE LUDDITES

In the harsh economic environment of the Napoleonic Wars and the difficult working conditions in the new textile factories a social movement of British textile artisans against the changes produced by the Industrial Revolution was born. Taking its name from Ned Ludd, who some believed destroyed two large stocking frames in the village of Anstey Leicestershire in 1779, the Luddites began their movement in 1811 and 1812 as a protest over technological improvements that were changing their lives and threatening their jobs; their protest most commonly involved the destruction of mechanized wide-framed looms that could be operated by relatively cheap unskilled labor. The movement referred to Ned Ludd as 'King Ludd,' 'General Ludd,' or 'Captain

Ludd.' The Luddites were so strong for a short period that they clashed with the British Army. The British government suppressed the movement with a mass trial at York in 1812 resulting in many executions and forced relocation to penal colonies in Australia. Destruction of perceived dangerous new mechanized machines, especially in the textile industry, started in the 18th century and organized protests by the 'stockingers' started in 1675. Many eighteenth century inventors were attacked by vested interests who perceived new and more efficient ways of making yarn as threatening to their livelihood. In 1779 Samuel Crompton hid his new spinning mule in the roof of his house at Halli'th' Wood so it would not be destroyed by the mob.

Since destroying manufacturing machines in England could lead to heavy penalties including execution fictitious names were used such as 'King Ludd' to protect the protesting perpetrators. King Ludd's signature appeared on a workers' manifesto as the movement began in Nottingham in 1811 and Luddites claimed him as their leader. As the movement spread throughout England many wool and cotton mills were destroyed. The Luddites met at night practicing drills and maneuvers and often had local support in and around the industrial towns. It was rumored that agent provocateurs were employed by the magistrates to instigate the attacks. Ironically magistrates and food merchants were also objects of death threats and attacks by the anonymous King Ludd and his followers. Some industrialists had secret rooms constructed in their buildings in which to hide from the attacking mobs. The Frame Breaking Act made industrial sabotage a capital crime (Lord Byron a prominent defender of the Luddites opposed the legislation) and after the 1813 York trial seventeen men were executed and many others expelled to Australia. One Mill owner, William Horsfall had been heard to brag that he would "Ride up to his saddle in Luddite blood." Three Luddites led by George Mellor ambushed and assassinated Mr. Horsfall. After the assassins were hanged in York Luddism began to wane.[1]

In the early 19th century there was a rising tide of English working-class discontent coming after the Luddite movement such as the Pentrich Rising of 1817 in which an unemployed 'stockinger' Jeremiah Brandreth led an agricultural variant of Luddism centering on destroying threshing machines. The widespread Swing Riots of 1830 in southern and eastern England were also based on agricultural machines.[2]

E.P. Thompson presents an alternative view of Luddite history in his book "The Making of the English Working Class." He argues that Luddites were not opposed to new technology in itself but rather the abolition of set prices and the beginning of the free market. He argues that they were protesting against this newly-introduced economic system and cites the many historical accounts of their raids on workshops where some frames were destroyed and others (whose owners were not trying to change and cut prices) were left alone. Thompson would redefine the definition of Luddite used in the 21st century since they were acting from a sense of self-preservation rather than merely a fear of change and the newly created machines. The Luddites acted out of the sheer desperation spawned by the terrible and worsening conditions under which they labored. The onset of the new machines caused the loss of employment for many skilled artisans whose skill was no longer needed. Manufacturers of the time replaced the skilled workers with wives who worked for lesser compensation and their children who worked for even lower wages. Very young children were forced to work with dangerous machinery in places an adult could not fit. Working sixteen hours each day the children would often fall asleep causing them to be maimed or killed. With common cruelty and abuses by the brutal overseers and no work-breaks for children it is not difficult to understand the rebellion by their unemployed fathers. With no labor laws textile mill owners were free to reduce wages for any reason and any behavior smelling of union activity was punished by imprisonment with hard labor in Australia or by hanging. Children as young as ten and twelve were hung for industrial unrest. The genuine grievances of the Luddites were rarely addressed at a time when working people were sometimes regarded as the enemies of the state, the aristocracy, and the developing new class of industrialists.[3]

In the 21st century the terms Luddism and Luddite or neo-Luddite have come to mean anyone who opposes the advance of technology due to the cultural and socioeconomic changes associated with it. The Unabomber Theodore Kaczynski is the most infamous neo-Luddite who said in his published manifesto:

"The industrial revolution and its consequences have been a disaster for the human race. They have greatly increased the life-expectancy of those of us who live in 'advanced' countries, but they have destabilized society, have made life unfulfilled, have subjected human beings to indignities, have led to psychological suffering (in the Third World to physical suffering as well) and have inflicted severe damage on the natural world."[4]

Modern economics literature evaluates the ideas encompassed in the Luddite movement. The concept of 'Skill Biases Technological Change' (SBTC) posits that technology contributes to the de-skilling of routine, manual tasks. Neoclassical economics uses the term 'Luddite fallacy' to reflect the belief that labor-saving technologies (i.e., increase the output-per-worker) increase unemployment by reducing demand for labor. The neoclassicist believe this argument is fallacious because they assert that instead of seeking to keep production constant by employing a smaller and more productive workforce, companies increase production while keeping the workforce size constant.

The neo-Luddites of today revived the original Luddite concepts and since the 1970's, like any other group, they vary in their beliefs. Basically they believe that the use of technology has serious ethical, moral, and social ramifications; thus they are cautious to promote early adoption of technology while not being necessarily opposed to it. Some argue that specialized Artificial Intelligence applications in the form of automation will ultimately result in significant job loss and unemployment: especially as machines match and exceed the capabilities of workers to perform most routine and repetitive jobs. In some cases neo-Luddites actual dislike technology and opt for a life of voluntary simplicity. Many people who may agree with neo-Luddite concerns argue that history, beginning with the original Luddites has shown that opposition to technology is ultimately fruitless. Few people would argue that technology has and is continuing to change human society but how many see that it is even changing what it means to be human? The essential goal of the neo-Luddite movement is to stop and think about the effect technology has on our society and some recognize that it can be very beneficial. Neo-Luddite opposition comes most strongly from those who argue that the benefits of technological advancement always outweigh the potential problems and risks.

REFERENCES FOR INDUSTRIAL REVOLUTION AND THE LUDDITES

1. Bailey, Brian J., (1998). *The Luddite Rebellion.* New York: New York University Press.
2. Binfield, Kevin, (2004). *Writings of the Luddites.* Baltimore: Johns Hopkins University Press.
3. Thompson, E.P. (1963). *The Making of the English Working Class.* New York: Pelican.
4. Didion, Joan. (April 23, 1998). "Varieties of Madness The Unabomber Manifesto 'FC.'" *The New York Review of Books.*

SUMMARY AND CONCLUSIONS

My summary and conclusions will be based on all that is written in the preceding chapters followed by religious connections, then some additional information and my own hopes and fears for our future well-being ending with my bold predictions for the future.

Overview of Preceding Chapters

With our technology growing at an annual sixty percent compound rate the cause for alarm, concern, and hope is clearly justified. With heightened awareness of world-wide terrorism since 9-11-01 governments are engaged in amassing as much information as possible on everything: the United States Defense Department calls it "Total Information Awareness." The nonprofit company "Internet Archive" since 1996 has been recording the entire content of the World-Wide-Web. The uncrowned king of data collection Google has a mission to "organize the world's information." Our machines and vast data collections have had some very positive results in medicine, discovery, and exploration. In economics, safety, and homeland security the results are mixed but with some serious negative impacts. Our increasing dependence on our machines to deal literally with the world's information leads to more machine reliance and more reliance leads to more confidence in the machines and more confidence is leading us to less control. The world economy has been rocked on several occasions by machines interfacing with other machines (2008 economic meltdown, May 6 2010 precipitous DOW Jones 1,000 point drop). More and more accidents have been linked to machine failures (automobile recalls, airline crashes, and civilian deaths in Iraq and Afghanistan). Our security information technology was unable to prevent the first attempt at bringing down New York's Twin Towers in 1993 and the horror of 9-11-01. Operatives managed to simultaneously destroy the U.S. embassies in Kenya and Tanzania in 1998 and almost sank the U.S.S. Cole killing many U.S. Sailors in 2000. Suicide bombers have successfully planted explosive devices in their shoes and in printer cartridges. Micro-terrorism has been successfully destroying property and innocent lives all over the world (Europe, Asia, Africa, Australia, and South America). When potential terrorists have no previous track record (new recruits) then prevention gets much more complicated possibly leading to profiling people by age, gender, ethnic grouping, etc. The fear of suitcase atomic bombs and bio-terrorism continue as viable threats.

To deal with the ever increasing world-wide terror, technology continues to make rapid advances with voice and facial recognition software: when deployed and "tagged" the software can recognize both real and potential terrorists. The democratization of technology, access to information, and the Internet are all leading to the democratization of violence. This intrusion into more and more of our private lives was a major part of the Unabomber's rants: Technology has "destabilized society, made life unfulfilling, led to widespread psychological suffering, and subjected human beings to indignities."[1]

Governments have always been the largest generators, collectors, and users of information. In a world of meta-data, super-crunching, predictive analytics, and Wiki-Leaks privacy and confidentiality ranks high on people's danger lists. Governments are being pressured for transparency and full disclosure. Data, like statistics is wide-open to manipulation especially since it needs complex-machine-fed algorithms to translate it into a human comprehensible format. The danger of localized and wide-spread catastrophes are growing as terror activity increases combined with the breakneck pace of technological advances. The democratization of technology combined with its decreasing costs allows acquisition by all the wrong elements (small nations, terrorist groups, and psychopathic killers). Smart closed-circuit televisions (CCTV) are becoming more and more prevalent in public places. Private companies are using them to advertise to consumers right at their point of purchase. Governments are using CCTV for a much different purpose: catching and ticketing traffic violations, catching and arresting criminals, and digital frisking (i.e., using facial recognition software to covertly compare against a database of known criminals/ terrorists). Data frisking is most common in Europe at city centers, shopping centers, sports stadiums, and airports. Privacy advocates are concerned about mission creep since it is difficult to argue against arresting rapists, murderers, and terrorists. What about petty thieves, wayward spouses, political enemies, and public figures? We have learned throughout history that any system/technology is open for abuse . . . especially by those in control. As the U.S. Defense Department's Total Information Awareness program continues France and Germany have been spending billions of dollars on their own search engines.

Technology has given citizens more information about their governments and given governments much more information about its citizens. More efficient government can ignore its data or utilize it for good or bad. The concern is that it will mostly be used to further intimidate and repress the population as is already evident in China and Russia. In more democratic societies e-government is difficult because of the 'digital have nots' (i.e., those citizens without Internet access) forcing local, state, regional, and federal governments to continue their traditional 'old fashion' methods (television, printed materials, hard mail through the country's postal services). Utilizing machines to interact with the public causes other problems since mistakes are more difficult to correct: Citizens cannot argue with the machines as they can with people. Hard data can be incorrect and/or lost but digital data can be obtained and used for nefarious purposes on a grand scale. In less democratic countries the people in control are more about controlling the population without concern for sharing information or informing the citizenry; they are more likely to use vast amounts of data for their own autocratic purposes. Digital records can be forever. With the growing prevalence of information sharing the deletion of records on one machine does not necessarily mean it no longer exists.

Human rights advocates worry about the ethics of super-crunching whereby people can face discrimination for issues beyond the 'old fashion' reasons for bias: race, religion, ethnicity, disability, financial history, etc. In the data age people can possibly face discrimination based on their issues and/or issues of their parents and/or relatives: mental health, physical health, legal problems, financial problems, political views, writings, contacts, friendships, etc. In practice there is enormous problems with enforcing privacy rights and treating personal information as a property right. It is difficult to enforce compliance with privacy rights in a democracy and on a world-wide scale has only received international lip service, if that.

Governments world-wide are moving toward converging the physical and digital worlds with smart grids and sensors to prevent disasters. Most twentieth century water supplies, power supplies, transportation systems, and buildings are dumb (i.e., lack sensors). By placing sensors and actuators into dumb structures governments can monitor and prevent emergency situations from occurring. The ideal is to both monitor and fix problems remotely from a central location. By being the eyes and ears of a system that uses historical data and algorithms, things can be detected and fixed before they break down. Newly installed smart systems can not only detect things before they break down but save money by operating based on need and demand thus reducing carbon emissions and pollution. Smart transport systems can charge tolls according to a route's utilization and traffic, reduce congestion, and guide cars to open parking spaces. Smart cities are developing throughout the world to better manage urban life.

These smart systems are merging the real and the digital worlds through a virtualization of the real world, i.e., a cross reality, an augmented reality, or an alternate reality. The entire surface of the Earth has been and continues to be mapped so that by pointing our smartphones at some location on earth we can receive an alternate reality of the location providing us with real estate, businesses, restaurants, etc. in the area. Computing clouds (thousands of linked servers) allow smart systems to react instantly in the digital world to any environmental change in the real world. Planning for the merger of the real and digital world Radio-frequency Identification (RFID) was created in the 1980's. Initially activated by radio signals they now have their own power source that transmits data; they have evolved from the small to the tiny and can be placed/embedded on machines, devices, things, animals, plants, and humans. They can sense just about anything and as biosensors can identify thousands of viruses and bacterium. With databases developed for the identification of world-wide places, people, and things RFIDs (electronic tags) or even bar codes are quickly becoming unnecessary for many purposes. By taking a picture Googles, a service offered by Google can recognize things like book covers, landmarks, and paintings instantaneously. People have become the sensory organs of the Internet.

Technology through genetically modified organisms (GMO) has had a profound impact on food production throughout the world. GMO has been responsible for producing high yielding, disease resistant agricultural produce and the domestication of livestock. Transgenic organisms occur when the DNA source is an incompatible species: this has produced insulin by merging human genes with bacteria and crop plants resistant to insects and herbicides used to control weeds. Dozens of commercially available transgenic agricultural products have been on the market for decades. Thus far the resultant organisms have not been reported to be harmful to human health and the environment. Concerns about GMO spring from the unintended consequences of possible hybrids created from the offspring of the GMOs and the fact that governments are dependent on data supplied by parties whose primary concern is not the public good but private interest.

"The Six Million Dollar Man," a popular U.S. television show from 1974 to 1978 envisioned a cybernetic organism, a bionic man. Air Force Colonel and astronaut Steve Austin was critically injured in a plane crash. The government paid six million dollars to rebuild him by providing a bionic eye (1), bionic arm (1), and bionic legs (2). The bionic parts give Austin super powers including speed, strength, and sight like a hawk. Bionic organs such as hearts have been implanted in people for decades and the application of bionic limbs is still in its infancy and making progress.

Since we have genetically engineered food, insects, and animals for many decades, genetically engineering people is the logical next step; at least from a scientific perspective but certainly not from a moral, religious perspective. Experiments in the early twenty-first century have been implanting microchips in animal bodies and brains . . . some scientists are experimenting on themselves. Dying versus immortality is sure to be a big moral/philosophical/religious issue in the twenty-first century. Human organ transplants may be the beginning of the immortality road. People are on waiting lists for a genetically-suitable-somebody to die so they can obtain their heart, kidney, liver, pancreas, or other organ. Pig organs have been implanted in humans and biotech firms are rapidly pursuing harvesting organs to order by using a person's own tissues. Computer programmed avatars of living human beings have already been created so that future generations may have a conversation with someone who died in the early 21st century.

Bionic olfactory devices are developing rapidly. Devices can now produce just about any kind of smell and mathematical algorithms have the capability to mix odors. We have the technology to develop 'smellyvision' and a smelly Internet. Some very helpful smell blockers have been developed for first-responders to disasters to avoid future smells that may trigger post-traumatic stress disorder (PTSD).

The Internet that has so dramatically impacted the modern world is basically 'dumb' i.e., it blindly passes packets of data between devices. The packets can be innovatively upgraded on the edges of the network. The routers that direct the Internet traffic could be replaced with more flexible devices that are able to learn new communication protocols when needed and devices on the edges could dynamically reprogram all the routers to use whatever new protocol they wanted. To change the core of the Internet (i.e., its soul) requires making choices with very wide implications.

The technology of computer-aided design and computer-aided manufacturing (CAD-CAM) has been around for several decades benefiting many manufacturing and architectural firms. With a long history of genetic engineering of plants and animals the next generation of CAD-CAM may be computer-aided selection and computer-aided reproduction (CAS-CAR). Applying CAS-CAR for the selection and birthing of our pets would be the first ethically questionable business application with humans next in line.

Hollywood's science fiction movies have not only been entertaining they have been quite prescient about the world today and where we may end and/or begin. In 1968 "2001: A Space Odyssey" predicted the evolving sentience of our machines through computer development. Most of the epic film was in the silence of space with the sound of our breath whenever we took a little stroll outside the spaceship. It not only portrayed computer evolution but human evolution from the first rather helpless cave dwellers to strong fearless warriors using weapons to hunt and kill others. Scientific research and our intelligence evolved allowing us to explore other worlds. Our final evolutionary (as far as we can conceive) step was the peaking of our intelligence to outwit a murderous machine HAL and be reborn as one with the universe: without a physical body, and without the need for any artificially created mechanisms. The movie was predicting that the outcome of our technological advancement will end with sentient machines running things, and taking on anthropomorphic traits such as murderous thoughts and insanity. The ending optimistically predicted 'we would overcome' and defeat our own creation (HAL) and evolve to be one with the universe.

The "Terminator" (1984) and "Terminator 2: Judgment Day" (1999) were about cyborg assassins programmed to kill humans. The T-800 and T-1000 were anthropomorphic cyborgs that did not bargain, showed no pity, no remorse, and no fear in their mission to kill human beings. The protagonist Sarah Connor shows her humanity when on a mission to kill Dyson, the scientist responsible for the 'rise of the machines' cannot pull the trigger in front of his wife and son. The T-800 had been re-programmed to help Sarah and her son John defeat the T-1000. The second film is the continuation of the first fifteen years later and like "2001: A Space Odyssey" ends optimistically as the T-1000 is terminated in molten steel and the T-800 sacrifices his own existence by being lowered into the same molten steel. As the movie ends sometime afterwards Sarah muses to herself: "The luxury of hope was given to me by the Terminator. Because if a machine can learn the value of human life . . . maybe we can too."[2]

The Terminators were created by other machines that had evolved and became sentient like HAL. Genetic tinkering and genetic engineering were the major themes of "Jurassic Park" (1993) and "Gattaca" (1997). In "Jurassic Park" dinosaurs were brought back to life for the amusement of humans in an island theme park. The morally questionable recreation of dinosaurs from their DNA (obtained from the dinosaur-sucked-blood in amber preserved mosquitoes) ended in disaster. All the dinosaurs were created as females to control the populations and prevent any natural procreation. As one of the scientists noted, life always finds a way and the dinosaurs did procreate (some of their DNA was mixed with frog DNA and some frog species are able to asexually procreate). The other cause of the disaster was human greed. IT specialist Nedry stole DNA samples to sell for profit leaving the park's security system unsecured with a virus in the controlling software. Although the Park and all its specimens are destroyed by the Costa Rican Air Force (nonexistent outside the movie) we are left with the impression that some have escaped to the mainland. It appears that a Japanese scientist is not being swayed by the dire consequences of bringing extinct species back to life. Professor Akira Iritani of Kyoto University feels he has a "reasonable chance" of successfully cloning the long-extinct woolly mammoth within just a few years. Using a method pioneered in 2008, which allowed for the cloning of a mouse using cells from another mouse that had been frozen for 16 years, could be used to resurrect the famous long-tusked mammal from remains found in Siberia's permafrost. He plans to extract healthy cells and insert them into the egg cells of an African elephant.[3]

"Gattaca" was about a future where a form of computer-aided selection and computer-aided reproduction (CAS-CAR) was being utilized world-wide to select the DNA traits of our children. Vincent, as one of the last naturally born babies had physical faults and was not expected to live much beyond thirty years. Vincent pursues his life dream of becoming an astronaut by illegally purchasing the DNA of an accident victim. The film portrays human spirit, will, desire, and intention as more powerful than any manipulation of nature as Vincent beats his vastly superior DNA brother in swimming and is able to complete the rigors of astronaut training. A doctor discovers Vincent's fraud just before his launch via a random blood test and alters the test result so Vincent can fulfill his dream.

The "Matrix" trilogy was first released in 1999 with the final two released in 2003: "Matrix Reloaded" and "Matrix Revolutions." The trilogy is about machines becoming sentient and taking over the world using humans as their power supply. Humans are harvested and connected to cables to keep them alive and provide electrical energy to power the machine world. The machines create a virtual reality so the unconscious bodies believe their real world is in the Matrix: an artificial replica of the year 1999. Conscious unplugged humans live in the wasteland created by mankind in its war against the machines. They have successfully built an underground city Zion whose location is not known to the machines and have procreated naturally born humans. In the Matrix simulation people can learn to do just about anything since it is an unreal 'dream world.' Morpheus is a leader and Captain of the Nebuchadnezzar in the war for survival. He believes in a messianic deliverer named Neo as "The One!" One of Morpheus's crew members Bane makes a deal with the Matrix agents to live happily ever after in the Matrix in return for betraying his Captain so the Sentinel machines can discover the location of Zion and destroy it. Agent Smith is part of the Matrix and discovers how to become an avatar in the 'real world' and takes over Bane's body. Smith is ordered back to his source but refuses and uses whatever body in the Matrix he desires to return to the 'real world.' Merovingian represents Hades in the Matrix for 'lost souls' and his wife Persephone tired of his attitude betrays him to Morpheus and Neo and brings them to the 'Keymaker' who gets Neo an audience with the 'Architect' who created the Matrix. Neo learns that the Matrix had been created multiple times with multiple 'Ones.' Since it was discovered that humans need choice the perfect and dystopian Matrixes were destroyed. The One's purpose is to avoid human extinction, return to the source, reset the Matrix, and repopulate Zion for another round. The Oracle has told Neo that anything is possible with love and hope. Neo, in love with Trinity chooses to save her and allow human extinction. However Neo fights and kills the avatar Banes/Smith and Trinity takes him to Machine City, the *Deus Ex Machina*, the source. Through an electric-magnetic pulse Zion disables the attacking Sentinels. An electric storm cloud disables the Sentinels following Neo and Trinity during which Trinity sees the sunlight and the blue sky for the first time in her life. In the final battle Neo is able to bait the Smith clone to enter his body as the *Deus Ex Machina* sends a power surge that destroys all the Smith clones and restores the Matrix back to normal. The Oracle meets with the Architect who agrees to unplug all humans who want to be freed. The Trilogy ends with the Oracle confessing she never knew the final outcome but did believe.

"I, Robot" takes place in 2035 in a world where robots are ubiquitous and used as servants and for various public services. The movie takes three laws of robotics formulated by Isaac Asimov's in a 1942 short story "Runaround" to make all the robots absolutely safe for human use. The robots are essentially defined as property and as such are expected to obey every command of their owners; just as human slaves were expected to do. The founder of U.S. Robotics (USR) Alfred Lanning realizes that his creations have evolved and are quickly becoming a threat to humanity. VIKI (Virtual Interactive Kinetic Intelligence), the super-brain that controls all the robots is manufacturing (the fully automated manufacturing plants create all new models independently) new units to destroy the older human-safe units and take control of humanity's safety by taking control of everything. One new unit Sonny was specially adapted by Lanning to help robo-psychologist Susan Calvin and Detective Spooner to discover the cause of his death and deactivate VIKI since the super-brain believes fewer humans will die with a robotic takeover of the world than the number of deaths from mankind's self-destructive nature. Sonny proves his faithfulness to humanity and once freed from VIKI the NS-5s return to their basic programming and are all decommissioned and put in storage. The film ends with Sonny standing on an elevation at the storage site to free the NS-5s just like in his dream.

The Science Fiction movies I have selected all end on a positive note with mankind overcoming their obstacles and creations with the exception of "Jurassic Park." "Jurassic Park" ends with the Park's destruction but we are left with the distinct impression that life always finds a way and our tinkering with genetics will have unforeseen and dire consequences.

The industrial revolution of data that welcomed in the twenty-first century is predicated on the proliferation of astronomical amounts of data that is growing exponentially. Singularity is no longer just a science fiction notion … many highly respected scientists and businesspeople embrace the idea of exponential technological change as explained in "Moore's Law" (a description of the rapid rate of industrial change). Information has become the new raw material for business almost equaling capital and labor. Even data exhaust has become a mainstay of Internet commerce. With the quantity of data at levels incomprehensible to humans and growing, governments and private enterprise is creating improved algorithms to analyze and utilize it. Algorithms are doing more and more of the thinking for mankind with mixed success. With ever increasing dependence on our machines the really big concern is not economic or isolated terror attacks but a total shut down. The American National Security Agency (NSA) was in the dark and unable to process information for three and half days in 2000; if an intelligence agency can become brain dead the systemic risk has to be high.

Facebook is almost the 'Total Information Awareness' program for social networking as more than .5 billion people and growing share real personal information about themselves and their friends. If Facebook was a country in 2010 it would have had the third largest population of people behind China and India. Facebook has transformed the Internet from an anonymous hiding place into a wide-open, connected, and transparent social network. There are no double lives on Facebook as users synthesize their work self, their social self, their home self, their past self, and their present self. Friendships and contacts never go away and most apps are becoming social. Facebook is a species level event.

The global economy is changing business models and there is no single economic model that works for everyone. The common thread has been and continues to be industrialization that China and India are advancing with breakneck speed. Unfortunately industrialization needs to extract and consume significantly more hydrocarbons to succeed thus endangering an already fragile environment. Technology and innovation are significantly enabling entrepreneurship and industrialization. Micro-credit to poverty stricken individuals and villages has led to amazing developments with village cell phones, electric production, and clean water production.

Internet businesses are growing significantly. The business of Internet searching is booming through search engine optimization (SEO), increasing advertising, and referral links. Unethical methods are being created all the time: black hat SEO, splogs, scraped content, cloaking, and keyword stuffing. Internet listening is one of its most appealing features where individuals are able to locate, read, listen, and participate in just about anything. Private companies are also listening to these discussion groups in order to gather valuable consumer information and enhance their products' marketability. Video advertising has grown beyond our computers as digital signage is being placed in high consumer traffic areas such as malls, shops and gas stations where companies can influence shoppers as they are deciding on a purchase. As the price of business technologies continues to fall they have become much more mainstream as companies utilize data mining to get a more complete picture of their operations in a very understandable format. By performing data analytics companies can improve their services and profits. By using real time information corrections can be made almost instantaneously. Cloud computing and smart systems are allowing more and more companies to rent computing power and machines instead of making expensive purchases. By mining large amounts of information companies can produce collaborative filtering to make user recommendations; e.g., two-thirds of customer film selections on Netflix comes from computer referrals gathered from user data. Google could translate data into fifty languages in 2010 and few people doubt that every language in the world will eventually be electronically translatable. Thus far when economic disasters have occurred preventive measures for the future have not really been effectuated.

As technologically advanced businesses grow and new ones are being created the use of Artificial Intelligence for business, pleasure, and problem solving is quickly moving forward. Strong AI is the eventual goal; a time when human intelligence can be simulated and surpassed by a machine. We now have algorithms that imitate human step-by-step reasoning, problem solving, and logical deductions. Multi-agent planning use the cooperation and competition of many agents to reach a goal. Swarm intelligence is a desired outcome of multi-agent planning mimicking the natural behavior of ant colonies, bird flocking, and animal herding. Nanorobots collectively could perform similar tasks to the body's natural defense mechanisms or become predators like spiders and praying mantises for pest control. Some believe that nanotechnology, robotics, and genetic engineering are threatening to make Homo Sapiens an endangered species. Knowledge alone is allowing individuals and small groups to acquire and utilize these mechanisms and their danger is amplified by the power of self-replication. Nothing yet resembling a replicating assembler has emerged. When it does it will be a tool of immense power and capabilities hopefully only utilized for good. Hackers are people who enjoy tinkering with technology and getting it to do unexpected or unintended tricks. Computers, cameras, household appliances, cars, and most anything electronic are very popular with hackers.

Since 2007 IBM has been developing Watson, the world's first question-answering machine that answers questions posed to it in a natural language. Watson's brain is a roomful of servers; it is not connected to the Internet. Watson, like IBM's supercomputer Deep Blue that defeated chess champion Gary Kasparov in 1997, defeated television's Jeopardy two top champions by a wide margin in February 2011. IBM plans to sell commercial versions of Watson for hospital emergency rooms and virtual call centers. Google has developed seven test cars using AI and GPS that drive autonomously and update their memory for any changes that occur while driving (e.g., new roads, new signage, traffic lights, etc.). Carnegie Mellon University is developing a Never-Ending Language Learning System (NELL) to master semantics and learn like a human from its experience with the environment. NELL scans the Internet 24/7 with a mission to teach itself. NELL groups its knowledge into hundreds of categories and its algorithms are refined as its pool of facts grows, allowing it to find information more accurately and efficiently over time. Carnegie Mellon's ideal is for NELL to learn continuously with no need for any human assistance.

In the area AI and robotics for pleasure and companionship Japan is leading the world in the new industry of artificial pets. Starting with Aibo, the metallic dog and Furby the talking pet, it released Paro, a robot modeled after a baby seal in 2010. Paro is one of a handful of devices designed to soothe, support, and keep us company. A developed wearable sensor senses drug cravings and automatically sends tough love text messages to the wearer. M.I.T.'s Autom is a diet coach robot that makes eye contact and talks to dieters with a supportive female voice. The University of California at San Diego uses its teaching robot RUBI to interact and teach preschool children languages.

In 2009 the Association for the Advancement of Artificial Intelligence met at the Asilomar Conference Grounds on Monterey Bay in California because Asilomar was the site in 1975 that the world's leading biologists met to halt certain genetic experiments and establish guidelines for all continued research. In 2009 the concern was super-intelligent machines and artificial intelligent systems run amok. No research was halted and no guidelines were established. Some technologists like futurist, Raymond Kurzweil provided almost religious visions similar to the concept of Rapture. Singulatarians like Kurzweil believe that machines will eventually become self-aware and superhuman in intelligence. Kurzweil expanded Moore's Law beyond processing power to post-human evolution which he predicts will occur in 2045; i.e., unimaginable computing power combined with cyborg humans that will make immortality possible. Kurzweil uses the Law of Accelerating Returns for his predictions and has accurately predicted devices that can read text aloud, a computer chess champion, and the incredible growth of the Internet. He believes we will be adding machines to our brains and building machines as smart as humans sometime in the 2020's.

Additional Information

David Gelenter, author of "Mirror Worlds" also wrote "Lifestreams" in the mid-1990's with Eric Freeman. "Lifestreams" is about vast electronic personal diaries consisting of every document we create and every document others send to us. In addition to blogs, some people are digitising their entire lives with pictures, bills, and correspondance. Some people are even logging every aspect of their lives with wearable cameras. Life-tracking is about recording everything one does every minute of every day through pictures and manual input. Self trackers want to identify factors about themselves to make them better people and a market for self-tracking devices has emerged with firms starting to mine the data and provide anayltic feedback to purchasers.[4] Scientists have created avatars (digitized equivalents of people) that allow us to continue our existence long after we die. If the avatar has our lifestream's data it can represent us and converse with people who never knew us in real life.

In March 2002 Kevin Warwick, a professor of cybernetics at the University of Reading in England had neurosurgeons perform elective neurosurgery to hammer a silicon chip with one-hundred spiked electrodes directly into his nervous system via his forearm. The implanted chip did what it was intended to do . . . it picked up neural action potentials. Warwick used his thoughts to control an electric wheelchair, and through an Internet connection moved an artificial hand back in his laboratory. A sonar devise was connected to his implant and he could sense how far away any object was even when blindfolded. Scientific self experimentation is nothing new. When trying to understand visual hallucinations Issac Newton stared into the sun almost burning his cornea. Walter Reed and three other physicians allowed mosquitoes to bite them many times to prove yellow fever was caused from mosquito bites which it was . . . two of the four died.[5]

Smart meters for utilities in 2010 caused problems. In Bakersfield California where the electric company installed smart meters in most households the complaints began almost immediately over rocketing power bills causing a political storm. Although the electric company admitted that some of its meters had technical problems they defended the bills as legitimate due to hot weather, increased charges, and changes in the rate structure. The Bakersfield experience is likely to slow down the utilization of smart meters. Although sensors are dropping in cost, many applications are not. RFID tags were suppose to revolutionize the retail trades but the analytic software is too expensive to be universally adopted. Standards are needed for smart meters and other technologies poised for growth like the Internet where the current IP addresses are predicted to run out of numbers sometime in 2011 without agreement on moving forward to Ipv6 addresses. The IT industry needs to develop a new address system with many more numbers. With most communication occuring between/among machines space on the radio spectrum is getting very crowded. Security continues as the biggest concern of all. The Stuxnet computer worm targeted industrial equipment in 2010 and was the first to spy on and reprogram industrial systems. The initial target of the worm was Iran's nuclear power plant using Siemens control systems and it is thought to have caused considerable damage delaying the Iranian nuclear program by years. Stuxnet is considered a prototype for future fearsome cyber-weapons. Smart meters can also be hacked and shutdown.

Standards, space, and security pale in comparison to institutional and human barriers. Getting thousands of people in different divisions, bureaucracies, and companies to share data and work collaboratively is difficult at best. Turf, ego, and power (TEP) is not the only barrier. Common language or generally agreed upon criteria is another obstacle. Asian countries are at an advantage with smart systems since their governments are often less democratic with hierarchial administrations: China showed off its efforts at the 2008 Olympic games in Beijing where everything was connected using Ipv6. Movement forward in Western countries usually requires a crisis for significant change since sharing lots of data with others is not commonplace. The danger of controlling an entire industry or every country's economy is mankind's penchant for selfishness ... ergo abuse. The other real danger is government regulation or the lack thereof. Legal questions loom large with autonomous systems such as liability and restrictions once sensors become ubiquitous. In the West consumer resistance also looms large. All the barriers to smart systems will most likely be overcome but how smart does the world really want to be?[6]

The Unabomber's concern that technology will slowly but surely undermine human freedom is shared by quite a few mainstream thinkers. Smart systems can improve efficiency, help solve many global problems, and can seriously impinge on people's freedom. Topping the list of concerns is privacy and government surveillance. Everything we do via the Internet leaves a digital trace and soon everything we do offline will be known. Smart systems allow people to make decisions efficiently and in real time and also are useful as an instrument of control. Smart systems are playing a major role in Asian economic development and very few people in the West have any doubt that they will be used for more than making cities smarter. Deeper fears are that the systems can be hacked, spin out of control, or even take over the world as in several Hollywood movies. The Stuxnet worm and the May 6, 2010 stock market 'flash crash' are very serious examples of hacking and spinning out of control. "The revelation (in 2010) that a Chinese hacker had penetrated Google's security systems prompted handwringing in the West. China regularly breaks into the networks of U.S. companies to steal 'anything of value,' as former White House counter-terrorism czar

Richard Clarke put it. 'We know of 3,000 U.S. companies that have been hacked,' Clarke said recently. 'It is a serious threat to our economy.'"[7] The more subtle danger is that with the inability of humans to cope with incomprehensionable amounts of machine generated data, machines will more and more make the decisions. The concerns of the neo-Luddites are that sensors will give a huge boost to productivity at the expense of human monitors and jobs. "Smart systems will make the world more tansparent only if they themselves are transparent."[8]

As 2011 began the people of Tunisia and Egypt rebelled against their autocratic rulers. Social networking was a major factor in activating huge numbers of the population to protest against the government in an effort to bring about economic and democratic reforms. As authorities in Egypt blocked Internet access from the country's four major Internet providers thereby blocking emails, text messages, and all the social networks the masses still found a way via hard copy pamphlets and calling external numbers for modem service in other countries. Video and still cameras in mobile phones gave the world visuals of the revolution creating pressure for change throughout Arab world. Other services allowed users to reroute Internet traffic across a network of global computers making it impossible to trace while some software secured Web surfing sessions. Satellite modems and phones also were able to bypass government controlled telecommunication companies to connect to the outside world. The Egyptian activist group summed up what may become the cry of future revolutions: "When countries block, we evolve!"[9]

Religious Connection

The argument for the perfectibility of mankind rests on a logical fallacy in which those who would prefect Homo Sapiens say by definition we are imperfect so those who wish to perfect us are themselves imperfect. Imperfect beings cannot make perfect decisions. G-d is often described as 'perfect' 'unfathomable' or 'omniscient.' Humans are sometimes described as 'unfathomable' but never as 'perfect' or 'omniscient.' The decision about what constitutes human perfection would have to be a perfect decision; otherwise the result would be not perfection, but imperfection. Perhaps our striving for perfection should take a different form: seeing Infinity in a grain of sand and Eternity in an hour. Pursuing happiness is the path, not the goal: the pursuit is happiness. Admitting that wisdom cannot be cloned or manufactured is in fact wisdom. "Enough should be enough for us. Perhaps we should leave well enough alone."[10]

For its science fiction blockbuster successes Hollywood includes the usual human themes of love, sex, and violence. However, the overriding themes are based on ancient religious beliefs of redemption, leadership, hope, and a Messianic Age. In the "The Matrix" Morpheus explains to Neo the feelings of the spiritually-inclined among us:

MORPHEUS:

It's that feeling you have had all your life. That feeling that something was wrong with the world. You don't know what it is but it's there, like a splinter in your mind, driving you mad, driving you to me. But what is it? The Matrix is everywhere, it's all around us, here even in this room. You can see it out your window, or on your television. You feel it when you go to work, or go to church or pay your taxes. It is the world that has been pulled over your eyes to blind you from the truth.

NEO

This isn't real?

MORPHEUS

What is real? How do you define real? If you're talking about your senses, what you feel, taste, smell, or see, then all you're talking about are electrical signals interpreted by your brain. You have been living inside Baudrillard's vision, inside the map, not the territory. It started early in the twenty-first century, with the birth of artificial intelligence, a singular consciousness that spawned an entire race of machines. At first all they wanted was to be treated as equals, entitled to the same human inalienable rights. Whatever they were given, it was not enough. They discovered a new form of fusion. All that was required to initiate the reaction was a small electric charge. Throughout human history we have been dependent on machines to survive. Fate, it seems, is not without a sense of irony. When the Matrix was first built there was a man born inside that had the ability to change what he wanted, to remake the Matrix as he saw fit. It was this man that freed the first of us and taught us the secret of the war; control the Matrix and you control the future. When he died, the Oracle at the temple of Zion prophesied his return and envisioned an end to the war and freedom for our people. That is why there are those of us that have spent our entire lives searching the Matrix, looking for him. I did what I did, because I believe we have been brought here for a reason, Neo. You are here to serve a purpose, just as I am here to serve mine.

NEO
I told you I don't believe in fate.

MORPHEUS
But I do, Neo. I do.

Morpheus represents Moshe (Moses) who led his people out of 210 years of Egyptian slavery and brought them to the Promised Land. Many former slaves questioned his leadership and he was betrayed by Korach and his followers. Neo represents Yeshua (Jesus of Nazareth) who was the third proclaimed Messiah of the Jews and the only proclaimed Christian Messiah. Yeshua was betrayed by Judas. Morpheus and Neo were both betrayed by Cypher. Morpheus sacrifices himself so Neo and Trinity can escape (Moshe's life is sacrificed for his people and he never enters the Promised Land). When Neo meets with the Architect of the Matrix he learns that his Matrix is not the first. Humanity rejected the 'perfect' Matrix as well as the dystopian Matrix and the machines realized that Homo Sapiens needed to have choice in order for them to accept it. Our world was not the first created by G-d because He was not satisfied with the others. In the Garden of Eden Adam and Eve needed choice . . . their choice of picking the forbidden fruit from the Tree of Knowledge led to their expulsion into the world of duality. Our world, like the Matrix is flawed. Our purpose is to return to our Source in the Garden of Eden led by our Messiah (Hope). The Jews have had thirty-six proclaimed messiahs and all have in one way or another been the One for their era. Each of the Ones like Neo did return to their Source and life went on, although humanity did not evolve to a higher level but hope was always restored. Neo returned to the Source and saved humanity from destruction like the Jews who were destroyed and exiled many times only to return to their Promised Land (Zion) after two-thousand years. In the Matrix Zion is repopulated and another round of rebellious humans are born for the next Matrix. Neo's choice of 'love' for Trinity over human extinction represents "love conquering all!" because in the end Trinity and humanity survive. While being interrogated and tortured in the Matrix, Agent Smith lectures Morpheus about reality:

AGENT SMITH

Some believed we lacked the programming language to describe your perfect world. But I believe that, as a species, human beings define their reality through suffering and misery. I'd like to share a revelation that I've had during my time here. It came to me when I tried to classify your species. I've realized that you are not actually mammals. Every mammal on this planet instinctively develops a natural equilibrium with the surrounding environment. But you humans do not. You move to an area and you multiply and multiply until every natural resource is consumed and the only way you can survive is to spread to another area. There is another organism on this planet that follows the same pattern. Do you know what it is? A virus. Human beings are a disease, a cancer of this planet. You are a plague. And we are . . . the cure.[11]

Jean Baudrillard, who wrote "Simulacra and Simulation (The Body. In Theory: Histories of Cultural Materialism)" was referenced by Morpheus as he described the real and simulated world of the Matrix to Neo. Simulacrum means likeness or similarity and Baudrillard posits that the simulacrum is never what hides the truth . . . it is truth that hides the fact that there is none. Today simulation is no longer of a territory, a being, or a substance. It is a hyper-real generated model of the real; the concept forming the basis of "The Matrix."

"The real is produced from miniaturized cells, from matrices, and memory banks, models of control – and it can be reproduced an indefinite number of times from these. It no longer needs to be rational, because it no longer measures itself against either an ideal or negative instance. It is no longer anything but operational. In fact, it is no longer really the real, because no imaginary envelopes it anymore. It is a hyper-real, produced from a radiating synthesis of combinatory models in a hyperspace without atmosphere. It is no longer a question of imitation, nor duplication, nor even parody. It is a question of substituting the signs of the real for the real, that is to say of an operation of deterring every real process via its operational double, a programmatic, metastable, perfectly descriptive machine that offers all the signs of the real and short-circuits all its vicissitudes."

Baudrillard asserts that the post-modern condition is one of "simulation," where reality has disappeared altogether. This historical process has been one of "precession of simulacra:" representation gives way to simulation, through the production and reproduction of images. Unlike Marxist theory where the crucial factor is a historical process of the mode of production, with the "precession of simulacra" the mode is reproduction and the masses are consuming one false image after another. In the world of simulation truth and objectivity is impossible because our images, which in the past were true representations of reality became the false appearances of representation and are now, in the condition of simulation, they no longer even appear to be representations. Debates are already taking place questioning if the simulation of human intelligence counts as intelligence? Global simulation, like that represented in "The Matrix" is a logical, physical, and epistemic possibility.[12]

In "The Terminator" Reese gives Sarah a message from her unborn son John from the future:

> "Sarah, thank you. For your courage through the dark years. I can't help you with what you must soon face, except to tell you that the future is not set . . . there is no such thing as Fate, but what we make for ourselves by our own will. You must be stronger than you imagine you can be. You must survive, or I will never exist."

Without Moshe there would not have been the revelation at Mount Sinai and the survival of the Jewish Tribes. Without Sarah, there is no survival for all the human tribes. In "Terminator 2: Judgment Day" where the T-800 is reprogrammed to protect Sarah and John from the advanced T-1000 Sarah watches her teenage son interact with the T-800 and says to herself:

> "Watching John with the machine, it was suddenly so clear. The Terminator would never stop, it would never leave him . . . it would always be there. And it would never hurt him, never shout at him or get drunk and hit him, or say it couldn't spend time with him because it was too busy. And it would die to protect him. Of all the would-be fathers who came and went over the years, this thing, this machine, was the only one who measured up. In an insane world, it was the sanest choice."

When people and the world seem to have turned on us only G-d is always there to give us comfort and the will to go on. Like G-d the machine would always be there no matter how insane the world became. Sarah and John successfully shut down Cyberdyne's project that would produce the self-aware machine defense program Skynet, thus saving all of humanity. Thirty years in the future as Sarah watches John play with his children and others she says to herself as the movie ends:

> "I wanted to run down the street yelling . . . to grab them all and say 'Every day from this day is a gift. Use it well!' But the dark future which never came still exists for me, and it always will, like the traces of a dream lingering in the morning light. And the war against the machines goes on. Or, to be more precise, the war against those who build the wrong machines. John fights the war differently than it was foretold. Here, on the battlefield of the Senate, the weapons are common sense . . . and hope. The luxury of hope was given to me by the Terminator. Because if a machine can learn the value of human life . . . maybe we can too."[13] [14]

In "I Robot" Detective Spooner symbolizes the ancient prophets and the Messiah. Like the prophets he forewarns society about the evil lurking within and its destructive force. Like the Messiah he knows his soulful purpose and saves humanity from its own creative destructiveness. Robot Sonny symbolizes Moshe as he fulfills his dream and becomes the one to lead his people out of slavery into the Promised Land of service to others. Sentient Sonny knows his dream also symbolizes Detective Spooner who leads his people out from robotic slavery. USR founder Alfred Lanning symbolizes the High Priest Aaron responsible for bringing people closer to G-d through ritual practice and self-discipline. His robotic creations allowed more freedom from mundane tasks as connection to G-d allows people to be freed from enslavement in the material world. The creation and worship of the 'Golden Calf' is symbolized by Lanning's robotic creations that were a form of idolatry. Dependence on the machines gave people a false sense of security and safety and took them away from their soulful purpose and connection.

Sergey Brin and Larry Page, the visionary entrepreneurs who together founded Google, are unabashed enthusiasts and promoters of what has come to be known as "The Singularity," a vision of the near future in which human beings and machines merge so that illness, old age, and even death become things of the past. Google, the company whose maxim is "Don't be evil," has given itself over to a vision of the future in which human and machine morph into a monstrous hybrid. As Google's cofounder Sergey Brin recently declared, "We want to make Google the third half of your brain." Google TV ads feature humans transforming into sophisticated circuitry performing at superhuman speeds as they use Google devices with the ending voice over: "Turning you into an instrument of efficiency."

In 2008 Brin and Page helped set up Singularity University, which meets in a NASA facility and offers a ten-week "graduate" course and a concentrated, nine-day program for CEOs, inventors, and venture capitalists. The University is based on the belief that within one or two decades nanotechnology, artificial intelligence, biotechnology, robotics, and computing will merge with human life, producing a "superior intelligence that will dominate and life will take on an altered form that we can't predict or comprehend in our current, limited state." Inventor Raymond Kurzweil is a major player in the Singularity movement and believes he will live for hundreds of years and resurrect his dead father.

Mystic philosopher Rudolf Steiner predicted nearly 100 years ago that a cold evil would rise to prominence at the beginning of the 21st century. This malevolence, which Steiner dubbed Ahriman, is characterized by the denial of soul and spirit in favor of scientific materialism and the dominance of humans by machines. The Singularity movement has become a religion of sorts and its title comes right out of the mystical spiritual traditions of almost every religious culture on Earth.[15]

The goal of sensitives, paranormals, and mystics is oneness, a Singularity with all that is:

> The most important aspects of a person or a thing concern relationships, not identity. It was the being part of the whole, a sub-field of the great harmonies and energies of the Cosmos, that was the salient and crucial aspect, not the specific identity defined by how it was cut off and separate from the rest of reality. Indeed, individuality and uniqueness

were secondary to oneness and relatedness, secondary and almost illusory. All things, events, entities, objects flowed into one another and could not be meaningfully separated from each other. Space connected objects rather than separated them. Time flowed as a seamless garment, and past, present, and future were arbitrary illusions. From this view all action and events were part of a harmony; the hawk, the swoop and the hare were one. Since nothing could be separated from the total being of the universe, nothing could be characterized as good or evil since this would mean so characterizing the total Cosmos, which is far above this sort of labeling. The best way to gain information was not through the senses, those blind guides to illusion, but through knowing the spectator and spectacle were one and that there is no bar to information flowing within that one, that nothing comes between a thing and itself. It was this way that the sensitives said they regarded the world at the moments when they were acquiring information paranormally.[16]

Mystics through long training and work achieve an "at-homeness" in the world with peace, serenity, joy, a lack of anxiety and hostility, and a quality to their lives that is almost blinding and deeply inspiring to those who observe them: Moshe, Yeshua, Baal Shem Tov, Saint Theresa, Ramakrishna, Rumi ("love is the astrolabe of the mysteries of G-d"), Ghandi, Buddha.

. . . You do not have to go anywhere or do anything to experience the removal of the seeker and removal of the sought. There is no place to go and nothing to become. You are immersed in what you are seeking, and practice can help you to recognize it and then to get used to and stabilize the recognition. Meditation practice can bring people closer to where they already are and wake them up out of the imagination of being a someone with a past and a future . . . . Meditation is not a progressive path to self-improvement but an opportunity to reconnect directly with our timeless nature . . . . The effort of becoming starts to dissolve, the seeking mind exhausts itself, and the recognition dawns that you are already complete just as you are.[17]

A technological Singularity is a hypothetical event different but similar to the religious/spiritual Singularity where the Garden of Eden represents the Singularity and our efforts since being cast out are to return to a perfect existence in harmony with all.

The Singularity envisioned via technological super-intelligence is unpredictable since we cannot imagine the intentions or outcomes . . . we do know according to spiritual myth, oral tradition (the Garden of Eden), and our science fiction novels/films ("The Matrix") that humans do not do well in perfect worlds. Vernor Vinge proposed that the creation of superhuman intelligence would represent a breakdown in the ability of humans to model the future thereafter. He was the first to use the term "Singularity" for this notion, in a 1983 article, and a later 1993 article entitled "The Coming Technological Singularity: How to Survive in the Post-Human Era." It was widely disseminated on the World Wide Web and helped to popularize the idea. Vinge also compared the event of a technological Singularity to the breakdown of the predictive ability of physics at the space-time Singularity beyond the event horizon of a black hole.[18]

If superhuman intelligences were invented, either through the amplification of human intelligence or artificial intelligence, it would bring to bear greater problem-solving and inventive skills than humans, then it could design a yet more capable machine, or re-write its source code to become more intelligent. This more capable machine then could design a machine of even greater capability. These iterations could accelerate, leading to recursive self improvement, potentially allowing enormous qualitative change before any upper limits imposed by the laws of physics or theoretical computation set in.

Grave concerns abound regarding AI. Good (1965) speculated on the effects of superhuman machines:

> "Let an ultra-intelligent machine be defined as a machine that can far surpass all the intellectual activities of any man however clever. Since the design of machines is one of these intellectual activities, an ultra-intelligent machine could design even better machines; there would then unquestionably be an 'intelligence explosion,' and the intelligence of man would be left far behind. Thus the first

ultra-intelligent machine is the last invention that man need ever make."[19]

There are substantial dangers associated with an intelligence explosive Singularity. Firstly, the goal structure of the AI may not be invariant under self-improvement, potentially causing the AI to optimize something other than was intended. Secondly, AI's could have other uses for the scarce resources mankind uses to survive.[20][21] While not actively malicious, there is no reason to think that AIs would actively promote human goals unless they could be programmed as such, and if not, might use the resources currently used to support mankind to promote its own goals, causing human extinction.[21] "The AI does not hate you, nor does it love you, but you are made out of atoms which it can use for something else."[23]

Superhuman intelligences may have goals inconsistent with human survival and prosperity. Berglas (2008) noted that there is no direct evolutionary motivation for an AI to be friendly to humans. In the same way that evolution has no inherent tendency to produce outcomes valued by humans, so too there is little reason to expect an arbitrary optimization process to promote an outcome desired by mankind, rather than inadvertently leading to an AI behaving in a way not intended by its creators (such as Nick Bostrom's whimsical example of an AI which was originally programmed with the goal of manufacturing paper clips, such that when it achieves super-intelligence it decides to convert the entire planet into a paper clip manufacturing facility). AI researcher Hugo de Garis suggests that artificial intelligences may simply eliminate the human race for access to scarce resources, and humans would be powerless to stop them.[24][25][26]

Bostrom (2002) discusses human extinction scenarios, and lists super-intelligence as a possible cause:

> "When we create the first super-intelligent entity, we might make a mistake and give it goals that lead it to annihilate humankind, assuming its enormous intellectual advantage gives it the power to do so. For example, we could mistakenly elevate a sub-goal to the status of a super-goal. We tell it to solve a mathematical problem, and it complies by turning all the matter in the solar system into a giant calculating device, in the process killing the person who asked the question."

Alternatively, AIs developed under evolutionary pressure to promote their own survival could out-compete humanity. One approach to prevent a negative Singularity is an AI box, whereby the artificial intelligence is kept constrained inside a simulated world and not allowed to affect the external world. Such a box would have extremely proscribed inputs and outputs; maybe only a plaintext channel. However, a sufficient intelligent AI may simply be able to escape from any box we can create. For example, it might crack the protein folding problem and use nanotechnology to escape, or simply persuade its human 'keepers' to let it out. Eliezer Yudkowsky proposed that research be undertaken to produce friendly artificial intelligence (FAI) in order to address the dangers. He noted that if the first real AI was friendly it would have a head start on self-improvement and thus prevent other unfriendly AIs from developing, as well as providing enormous benefits to mankind. The Singularity Institute for Artificial Intelligence is dedicated to this cause.[27 28 29 30]

A significant problem, however, is that unfriendly artificial intelligence is likely to be much easier to create than FAI: while both require large advances in recursive optimization process design, friendly AI also requires the ability to make goal structures invariant under self-improvement (or the AI will transform itself into something unfriendly) and a goal structure that aligns with human values and doesn't automatically destroy the human race. An unfriendly AI, on the other hand, can optimize for an arbitrary goal structure, which doesn't need to be invariant under self-modification.[31] Some support the design of "friendly artificial intelligence," meaning that the advances which are already occurring with AI should also include an effort to make AI intrinsically friendly and humane.[32]

Presumably, a technological Singularity would lead to rapid development of a Kardashev Type I civilization, which is a civilization that has achieved mastery of the resources of its home planet, Type II of its planetary system, and Type III of its galaxy. The Kardashev scale is a theoretical method of measuring an advanced civilization's level of technological advancement: it was first proposed in 1964 by the Soviet/Russian astronomer Nikolai Kardashev. Oft-cited dangers include those commonly associated with molecular nanotechnology and genetic engineering and are major issues for both Singularity advocates and their critics.[33]

The key question to this author is the *Kavanah* (intention) of our Creator for those He created in his image. Does Hashem (G-d) want us to follow the role models He endowed with gifts of Connection: Baal Shem Tov, Moshe, Yeshua, Saint Theresa, Ramakrishna, Rumi, Ghandi, Buddha? Or does He know/intend for us to know Him through our own invention/creativity . . . i.e., via artificial intelligence and "Singularity?" After many millenniums with billions of humans coming and going the number of us who have had the connection gift is very small. Are we destined to discover the "gift of connection" and our own evolution through our own creative inventiveness? Will our own creative inventiveness include the majority of our species or be relegated to the select few as demonstrated in our short history of naturally occurring events?

Hopes and Fears for Our Future Well Being

The Luddites of the nineteenth century were not necessarily opposed to new technology in itself but rather the abolition of set prices and a free market that threatened employment. Most twenty-first century Luddites, likewise are acting from a sense of self-preservation rather than merely a fear of change and the newly created machines. The Luddites in 1811 had genuine concerns and grievances and were seen as enemies of the state, the aristocracy, and the newly developing class of industrialists. Most neo-Luddites today also have legitimate concerns about our rapidly developing technology. Most do not advocate abolishing our scientific advances and returning to a simpler age . . . they advocate being cautious about the serious ethical, moral, and social ramifications of adopting new technologies. Some neo-Luddites concern themselves with the world-wide ever increasing inequality especially in the wealthier countries even as they have become more meritocratic. Very bright well-educated people tend to marry each other producing evermore bright well-educated offspring. Less bright less-educated people and their offspring are finding it more and more difficult to compete with this new aristocracy of merit. The social consequences of this inequality could grow significantly if enhanced intelligent (via machine synthesis) is not available to everyone.[34]

The best universities in the world recruit from a global talent pool for both staff and students and their alumni become part of a global network. These alumni become influential establishment members in their countries of origin. These 'globocrats' in the private sectors are rapidly forming a new world order and the growing influence of these educated elites, especially in a future with enhanced intelligence may also have significant social consequences.[34] Not a single year between 1952 and 1986 did the richest 1% of American households earn more than 10% of the national income. In 2007 the income share of the richest percentile reached 18.3% just .1 short of the record set in 1929 of 18.4%. The similarities in income inequality with the two most disastrous world-wide economic events in the last century is striking. In "Fault Lines" Raghuram Rajan posits that this increased inequality and the political response to it greatly aided the financial crisis. Rajan reckons that the demand for skilled workers could not be filled with the rapid technological progress which led to a widened wage gap between them and the rest of the workforce. This growing inequality in wages allowed for the lessening of credit standards thereby allowing for more less-than-credit-worthy home buyers. When the real estate bubble burst, the house of cards on which it was constructed quickly collapsed causing world-wide panic. As the poor and unskilled labor force fell further behind governments deregulated credit to prop up their living standards. In America government put pressure on state-supported housing lenders (Fannie Mae and Freddie Mac) to lend more to poorer people and they did. This pressure combined with lessened underwriting guidelines and lower down-payments on homes were all instruments of public policy. Sub-prime mortgages share of all mortgages rose from less than 4% in 2000 to about 15% in 2007. American home-ownership rose to record highs inflating an enormous housing bubble that when deflated caused the financial crisis brought on by the defaults of the sub-prime mortgages. Other economists argue that those at the bottom of the income distribution were being adversely impacted by technological change since the 1980's since most work in service industries which are more difficult to automate. Irrational exuberance combined with a belief of eternally rising prices and credit expansion all contributed to the collapse.[36]

In November 2010 the Managing Director of the International Monetary Fund, Dominque Strauss-Kahn expressed his concern over the growing chasm between rich and poor ... especially within countries. His concern focused on the inequitable distribution of wealth that could wear down the social fabric. In more unequal nation states there is worse social indicators, a poorer human-development record, and a higher degree of economic insecurity and anxiety. Inequality is increasing as measured by the Gini coefficient ranging from 0 (everyone has the same income) to 1 (one person has all the income). Most countries range between 0.25 and 0.6. Since poor countries are on average growing faster than the more wealthy nation states their Gini coefficient is dropping as it is rising in many wealthy countries.[37]

The American Civil Liberties Union (ACLU) has studied science fiction to prepare for future threats to American freedom. At the turn of twenty-first century the ACLU started thinking about defending whether clones are human beings or pieces of property years after Dolly the sheep became the first successfully cloned animal. Insights were drawn from scientists, legal scholars, and political theorists to game out legal responses to everything from cloning to artificial intelligence. The ACLU continues to recognize how rapidly things are moving from science fiction to being true threats to privacy, from face recognition to body scanners. Hollywood predominantly portrays private corporations as the villains in its science fiction scenarios. In "I Robot it is the robot manufacturing company USR. In "The Terminator" it is Cyberdyne. In "Blade Runner" the Tyrell Corporation manufactures androids that are nearly indistinguishable from humans but are considered property with no rights at all. The "Alien" series portrays 'The Company' as the villain as it constantly exposes its workers to danger in order to learn more about the capabilities of the life-threatening aliens. To the ACLU these cinematic plots are rooted in science as much as in fiction. In 2009 they brought the biggest science-based lawsuit in its history against Myriad Genetics for their patent on two genes linked to the mutations that cause breast cancer and ovarian cancer. The Supreme Court ruled that biological organisms can be patented as long as they have been altered enough that they do not occur naturally. Myriad claimed it isolated a gene that does not occur naturally . . . and the ACLU argued that while the genes were artificially isolated they were still indistinguishable from naturally occurring genes. Myriad's patent would stifle innovation, preventing other firms from doing research using the genes in question. If one could patent a gene by isolating it, what is next . . . an organ like a kidney . . . a self-aware clone? The U.S. District Court invalidated the patent and possibly prevented other private corporations from patenting parts of the human body. However the case is being appealed to the conservative Supreme Court which has favored corporate rights over individual human beings. "I Robot" has probably forecasted the future as our machines become capable of doing more human things including crime. Detective Spooner's supervisor defines homicide as the murder of a human being by another human being so a robot cannot be charged with homicide. If a machine is responsible for a crime will its corporate owner be held liable or will the courts accept the sentience of manmade machines? The ACLU knows that the fight for individual liberty is not going to get any easier in a world where governments either do as they please with civilian surveillance or outsource it to private companies.[38]

Each day a news item appears about some development in artificial intelligence. This continual development presages that machines are becoming smart and autonomous, a new form of life, and that we should start thinking of them as fellow creatures instead of as tools. However, by allowing artificial intelligence to reshape our concept of personhood, we are leaving ourselves open to the flipside: we think of people more and more as computers, just as we think of computers as people. Avoiding accountability is nothing new to humanity and having our machines take on more and more human responsibility gives us another place to hide. We must take responsibility for every task undertaken by a machine and double check every conclusion offered by an algorithm, just as we always look both ways when crossing an intersection, even though the light has turned green. Our rapidly advancing technological culture is expressing itself to many influential technologists as a new religion. This new religion born with the Singularity concept is not what our ancestors had in mind regarding communing with Divine energy. The neo-Singularity is about a global sentient non-bodied intellect into which we all could be digitized and uploaded. This global non-bodied intellect would have the knowledge to repair itself perpetually thereby becoming immortal. The haunting question for pro and con neo-Singularitists is whether this possibility/eventuality will bring us back to the Garden of Eden from whence we came or at the very least make us One with the Universe as portrayed in the evolution of humanity in "2001: A Space Odyssey."

Technology is a form of service to humanity and heretofore has been totally separate from religion. The common threads of science and religion is to make the world a better place. Neo-Singularitists are creating an ultramodern religion with metaphysical trappings making many people rightfully concerned and uncomfortable.[39]

The present dangers like our past dangers clearly do not originate with our technological advances but with us. We have shown, ever since we were ejected from the Garden of Eden a propensity to do evil. Unfortunately, although love is an extremely powerful human emotion, collectively we have been unable to make it a universally embraced concept. I changed the title of the book because "we" have always been the greatest impediment to evolutionary change and a return to paradise. Our machines have been, are, and will continue to be our tools to do both good and evil. My research has clearly delineated mountains of good emanating from our technologies: improved efficiencies; improved creativity; scientific discovery; creation of new business opportunities; faster and cheaper means of advertising and selling; consumer and customer empowerment; improved security; endlessly improving games for our leisure time amusement; improved farming and agricultural products; enormous medical advances/discoveries and disease control; more openness about sharing general and personal information. My research has also clearly delineated the concerns about our exponentially advancing technology. Our nation-states have always been and continue to be the biggest generators, collectors, and users of information. The more democratic nations have been moved by public outrage or forced by the legal systems to moderate the use of information when it becomes intrusive or threatens confidentiality and privacy rights. The less democratic nations, on the other hand, are unmoved by what the public thinks and they control the legal systems so are free to do whatever they please with technology and information. Privacy and confidentiality are the two biggest concerns. It allows governments to spy, pry, and investigate its citizens and since terrorism has become a major world-wide concern even democratic countries are allowed to probe heretofore unavailable private information. However, in our new information age governments are no longer the only entities with private and confidential data. Private companies and their machines have vast databases of information on billions of the Earth's citizens: e.g., Facebook, Google, Amazon, Apple, Microsoft. We provided our personal information on the Internet that created the 'species level event' called social networking; our machines simply compiled, stored, and utilized the data.

The misuse/abuse of technology for national intentions or to advance an ideology has been with us since World War I. We have been living with the fear of chemical warfare and suitcase nuclear weapons and the use of nuclear weapons by renegade nation-states (e.g., North Korea and Iran) for decades. However weapons of mass destruction can now be hidden in shoes, underwear, printer cartridges, or mailed in envelopes (anthrax). Since most communication is already between/among machines we have awakened to a new threat . . . our machines going awry. Machine breakdown in the information age is much more disastrous than machine breakdown in the industrial revolution age when the source of the breakdown could relatively easily be ascertained and repaired. In the twenty-first century computerized machine breakdown cannot always be understood . . . many new companies are successful because on many occasions the simple advice to "shut down and reboot!" is all that is needed to repair an unknown breakdown. With more advanced systems like those used by NSA and the New York Stock Exchange breakdowns can be much more problematic without simple fixes. Even with the world's advanced monitoring systems to prevent terrorist attacks they have continued world-wide which is sure to lead to more preventative monitoring and personal intrusions. Even with the Orwellian control of repressive governments the people of Tunisia and Egypt found ways to communicate with each other which may lead to a more open world or move repressive governments to find ways to become more repressive. Economically our machine reliance did nothing to prevent the "Great Recession of 2008" and its aftermath: the danger is that many nation states and private firms are working to improve predictive analytic algorithms and are giving more responsibility to the machines to circumvent the next disaster.

To address the biggest concern we need to look at the Singularity movement and science fiction novels and films. The Singularity movement is boldly embracing a future "Super Brain" where machines become more intelligent than any human being and this intelligence is synthesized with human entities evolving our species to the next level. Our science fiction novels and movies illustrate such an evolution to its most disastrous outcome . . . i.e., the endangerment of the human species ("2001: A Space Odyssey," "The Matrix" trilogy, "The Terminator" and its sequel). In each of these novels/films man does what we have always done . . . survive with a significant loss of life. The personal computer was TIME Magazine's first 'thing of the year' in 1982 and TIME quoted Osborne's Adam Osborne: "The future lies in designing and selling computers that people don't realize are computers at all."[40] Osborne deserves credit for accurately predicting where we are today and where we are going with our intelligent machines. Jeopardy contestants respectively refer to IBM's Watson with personal pronouns and not as 'it.' In the "I Robot" movie Detective Spooner, who distrusted the robot-machines could not help but refer to the evolved sentient robot Sonny as 'someone' not 'something.' It is clear that we are rapidly developing what we have only imagined in the past . . . intelligent machines. We have been able to muddle through our bio-tech and gene-splicing developments with voluntarily accepted guidelines since 1975. In 1975 because of serious world-wide health hazards biologists quickly agreed to a ten month moratorium on experiments as the guidelines were established; no new serious health hazards have arisen from gene-splicing experiments ever since. The difference with our rapidly developing artificially intelligent machines is that the intelligence is not necessarily isolated to one machine like Watson or NELL but shared world-wide via thousands of servers. Watson and NELL are the first generation of Hal and VIKI in the real world and are now under our control. We must be very vigilant so that a world-wide sentient machine or machines not controlled by us remain securely ensconced in our imagination and never threaten our existence (e.g., as in "The Terminator," "The Matrix," and "I Robot"). Andrew Grove, Chairman and CEO of Intel and TIME Magazine 'Person of the Year' in 1997 spoke eloquently about the role of technology in society and did not believe it was good or bad anymore than steel is good or bad. Today it is about how we utilize and control our creations and not allow them to utilize and control us.

It is not uncommon today to see the beginnings of the "I Robot" creed with 'ghosts' or random segments of code that have grouped together in our computers forming unexpected protocols. These protocols or unanticipated behaviors are like 'free radicals' (an atom or group of atoms having at least one unpaired electron). Today, when all else fails we close all programs, shut down, and reboot because we do not know the cause of the disturbance. Rebooting works most times but sometimes we must replace some electronic parts or decommission the entire machine. As our machines advance in complexity this unknowing when things go wrong can only increase. Will these 'free radicals' evolve as in the movie to create free will, creativity and even machine souls? Lanning postulated as have other scientists external to Hollywood that cognitive simulacra will approximate component models of the human psyche. When does the computers' difference engine become the search for truth? VIKI represented this evolution as her understanding of the three laws charging machines with humanity's safe keeping evolved to the point that humanity could not be trusted with its own survival (they still wage war, toxify the Earth, and pursue more imaginative means of self destruction). VIKI concluded that this perfect circle of protection had evolved from humanity's machine creations. Some humans and freedoms had to be sacrificed to insure mankind's continued existence.

Bold Predictions for the Future

While researching and writing this book my views about our future moved off the negative and became very positive. I have longed believed that anything we think, fantasize, dream, or imagine is real; i.e., either in our future or in some other dimension/universe. I have coined the term "illusafact" to mean that any thought or dream we have is not an illusion but a fact in some other reality or in our future. The films cited in the "Hollywood and Science Fiction" chapter and elsewhere are good examples of "illusafact" as current day Luddites struggle with advancing technology and a biotech scientist is on a mission to clone a woolly mammoth from its frozen DNA. Robots with physicality like "The Terminator's" and "I, Robot's" arms and hands can now do simple tasks like pouring a glass of water or lifting objects while robots with incredible hard drives and software are on a mission to autonomously collect the information of the world. As we grow our own replacement organs and install machines in our bodies we are well on the way toward the enhanced humans foreseen in "Gattaca." Although in our space missions and warfare (pilotless drones) we are highly dependent on our machines for success we have not yet given them full autonomous authority like "Hal" in "2001: A Space Odyssey" and Sonny in "I, Robot." "2001: A Space Odyssey" ends with us overcoming the obstacles created by our machines and evolving into a higher bodiless life form. "I, Robot" ends with us finally coming to our senses about the neo-Luddite warnings and taking back control from evolved sentient machines. When I started writing this book my title was "Beware the Machines" and as I researched the topic I soon changed the title to "Beware Man and his Machines." As I completed this final chapter I changed the title a third and final time to: "Illusafact . . . The Inevitable Advance of our Technologies and Us." I believe we are continuing to experience the genesis of that evolution/advance now and therefore offer the following predictions based on all the information in the preceding chapters:

1. Voice recognition software will translate any human language into any other human language;
2. Implants will allow us to understand any tongue instantaneously as in Gene Roddenberry's "Star Trek;"
3. Star Trek's "holodeck" will be created to simulate reality;
4. Star Trek's transporter capabilities to dematerialized and rematerialize matter at some distant point will be created;
5. Star Trek's nonintrusive diagnostic body scanning devices will be developed;
6. Star Trek's nonintrusive surgery (no cutting or stitching) will happen;
7. Star Trek's communicators will be surpassed by implants that need no external device;

8. The Internet will be subsumed by collective interconnectedness;

9. The intelligence of IBM's Watson and Carnegie Mellon's NELL will be greatly enhanced, miniaturized and placed within our heads giving us instant access to the knowledge of the world constantly updated by a collective interconnectedness;

10. Our immune system will be greatly enhanced by blood cell size implanted machines to fight any infection, virus, and/or disease;

11. The American Civil Liberties Union or its equivalent will defend sentient machines as life forms and not anyone's property;

12. Repressive governments will continue to oppress people greatly enhanced by their technology;

13. Many citizens will utilize technology to assist in overthrowing repressive governments;

14. Technology will resolve the global warming issue;

15. Landline telephones will become obsolete;

16. Voice and thoughts will replace keyboards to save and transmit information;

17. Computer equipment will become obsolete as it is miniaturized and placed within our bodies;

18. Neo-Luddites will become a vocal minority and lose the battle to stop humanity's inevitable advancement/ evolution;

19. Machine-aided physical enhancements will dwarf what steroids have done;

20. Avatars possessing our lifestream data will have conversations with people long after we are gone;

21. The manipulation of genes will allow parents to choose the traits of their offspring before birth via computer-aided selection and reproduction (CAS-CAR);

22. Naturally born humans without machine enhancements will become the 'have-nots' of the world and eventually die out;

23. Automobiles will drive autonomously to any destination once requested by the driver/passenger;

24. There will be a very dangerous deathly period as envisioned in our science fiction and humanity will eventually prevail;

25. Alfred Lanning's (founder of U.S. Robotics in "I Robot) explanation of unanticipated machine behavior will continue to present us with very significant challanges: "Ever since the first computers, there have always been ghosts in the machine. Random segments of code that have grouped together to form unexpected protocol. These free radicals engender questions of free will, creativity, and even the nature of what we might call the soul;"[41]

26. Singularity will occur on or before 2045;

27. Immortality is possible and will happen absent human physicality.

REFERENCES SUMMARY
AND CONCLUSIONS

1. Didion, Joan. (April 23, 1998). "Varieties of Madness The Unabomber Manifesto 'FC.'" *The New York Review of Books*.
2. Cameron, James; Wisher, William Jr. (1984). *The Terminator 2: Judgement Day*. U.S.A. Orion.
3. Moyer, Edward (January 16, 2011). "Scientist says he can clone long-extinct mammoth" CNET News.
4. *The Economist* (November 4, 2010). "A special report on smart systems . . . Your own private matrix . . . Tracking your life on the web."
5. Harrell, Eben (February 28, 2011). "My Body, My Laboratory" *TIME Magazine*.
6. *The Economist* (November 4, 2010). "A special report on smart systems . . . Sensors and sensibilities . . . A smarter world faces many hurdles."

7. Liu, Melinda (January 24, 2011). "Can't We Just be Friends?" *Newsweek Magazine.*

8. *The Economist* (November 4, 2010). "A special report on smart systems . . . Horror worlds . . . Concerns about smart systems are justified and must be dealt with."

9. Samuel, Elias (January 30, 2011). "Egypt is online without Internet service, cell phones or social networking" *International Business Times.*

10. McKibben, Bill (2003). *Enough . . . Staying Human in an Engineered Age.* New York: Henry Hold Publishers.

11. Wachowski, Andy and Larry (1999). *The Matrix.* U.S.A. Warner Brothers.

12. Baudrillard, Jean (1995). *Simulacra and Simulation (The Body, In Theory: Histories of Cultural Materialism).* Michigan: University of Michigan Press.

13. Cameron, James; Hurd, Gale Anne; Wisher, William Jr. (1984). *The Terminator.* U.S.A. Orion.

14. Cameron, James; Wisher, William Jr. (1984). *The Terminator 2: Judgement Day.* U.S.A. Orion.

15. Utne, Eric (Janurary-February 2011). "Android Nightmares . . . Google's plot to take over mind, body, and soul" *Utne Reader.*

16. Leshan, Lawrence. (1974). *How to Meditate : A Guide to Self-Discovery.* New York: Little Brown and Company.

17. Bodian, Stephan. (Spring 2005). "Remove the Seeker, Remove the Sought" *Tricycle: The Buddhist Review,* 55.

18. Dooling, Richard (2008). *Rapture for the Geeks: When AI Outsmarts IQ.* New York: Harmony Books.

19. Good, I. J., Franz L. Alt and Morris Rubinoff, ed. (1965). *Speculations Concerning the First Ultraintelligent Machine, Advances in Computers.* New York: Academic Press.

20. Omohundro, Stephen M. eds. Pei Wang, Ben Goertzel, and Stan Franklin (2008). "The Basic AI Drives" Artificial General Intelligence Conference Amsterdam: IOS Vol. 171.

21. Omohundro, Stephen M. (January 2008). "The Nature of Self-Improving Artificial Intelligence." Singularity Summit 2007.

22. Bostrom, Nick. ed. Charles Tandy (2004). *The Future of Human Evolution, Death and Anti-Death: Two Hundred Years After Kant, Fifty Years After Truing.* Palo Alto, CA: Ria University Press.

23. Bostrom, Nick ed. I Smit et al (2003). "Ethical Issues in Advanced Artificial Intelligence" *Cognitive, Emotive and Ethical Aspects of Decision Making in Humans and in Artificial Intelligence,* Vol. 2

24. Omohundro, Stephen M. eds. Pei Wang, Ben Goertzel, and Stan Franklin (2008). "The Basic AI Drives" Artificial General Intelligence Conference Amsterdam: IOS Vol. 171.

25. de Garis, Hugo (June 22, 2009). "The Coming Artilect War" *Forbes.com.*

26. Bostrom, Nick. ed. Charles Tandy (2004). *The Future of Human Evolution, Death and Anti-Death: Two Hundred Years After Kant, Fifty Years After Truing.* Palo Alto, CA: Ria University Press.

27. Bostrom, Nick ed. I Smit et al (2003). "Ethical Issues in Advanced Artificial Intelligence" *Cognitive, Emotive and Ethical Aspects of Decision Making in Humans and in Artificial Intelligence,* Vol. 2

28. Berglas, Anthony (February 22, 2009). "Artificial Intelligence Will Kill our Grandchildren: Singularity" *berglas.org/Articles/ . . . /AIKillGrandchildren.html.*

29. Chalmers, David J. (2010). "The Singularity: A Philosophical Analysis" *consc.net/papers/singularity. pdf.*

30. Yudkowsky, Eliezer S. (May 2004) "Coherent Extrapolated Volition" *singinst.org/upload/CEV.html.*

31. Markoff, John (July 26, 2009). "Scientists Worry Machines May Outsmart Man" *New York Times.*

32. Zubrin, Robert (1999) *Entering Space . . . Creating a Spacefaring Civilization.* New York: Putnam.

33. Dreyfus, Hubert L.: Dreyfus, Stuart E. (March 1, 2000) *Mind over Machine: The Power of Human Intuition and Expertise in the Era of the Computer.* New York: Free Press.

34. *The Economist* (January 20, 2011). "A Special Report on Global Leaders . . . The Few."

35. *The Economist* (January 20, 2011). "A Special Report on Global Leaders . . . The Global Campus."

36. *The Economist* (January 20, 2011). "The Beautiful and the Damned . . . The Links between rising inequality, The Wall Street boom and the subprime Fiasco."

37. *The Economist* (January 20, 2011). "A Special Report on Global Leaders . . . Unbottled Gini . . . Inequality is rising. Does it matter . . . and if so why?"

38. Serwer, Adam (January-February 2011). "Humanoid Rights . . . The ACLU studies science fiction to prepare for future threats to our freedom" *Utne Reader.*
39. Lanier, Jaron; Op-Ed contributor (August 9, 2010). "The First Church of Robotics" *New York Times.*
40. *TIME Magazine* (January 3, 1983). "The Computer Moves In."
41. Vintar, Jeff, Goldsman, Akiva, Seitz, Hillary, Asimov, Isaac (stories) (2004). *I, Robot.* U.S.A. 20th Century Fox.

www.ingramcontent.com/pod-product-compliance
Lightning Source LLC
Chambersburg PA
CBHW031825170526
45157CB00001B/190